People and Things

James M. Skibo • Michael Brian Schiffer

People and Things

A Behavioral Approach to Material Culture

James M. Skibo
Illinois State University
Normal, IL
USA
jmskibo@ilstu.edu

Michael Brian Schiffer
University of Arizona
Tucson, AZ
USA
schiffer@email.arizona.edu

The following chapters are reprinted in modified form with permission from the indicated sources: ***Chapter 3:*** Exploring the Origins of Pottery on the Colorado Plateau (with Eric Blinman), *Pottery and People* (1999), edited by J. M. Skibo and G. Feinman, University of Utah Press, Salt Lake City. ***Chapter 4:*** Smudge Pits and Hide Smoking Revisited (with John G. Franzen and Eric C. Drake), *Archaeological Anthropology: Perspectives on Method and Theory* (2007), edited by J. M. Skibo, M. Graves, and M. Stark, University of Arizona Press, Tucson. ***Chapter 5:*** The Devil Is in the Details: The Cascade Model of Invention Processes, *American Antiquity* (2005) 70: 485-502. ***Chapter 6:*** *Ball Courts and Ritual Performance* (with William H. Walker), *The Joyce Well Site*, edited by J. M. Skibo, E. McCluney, and W. Walker, University of Utah Press, Salt Lake City. ***Chapter 7:*** Social Theory and History in Behavioral Archaeology, *Expanding Archaeology* (1999), edited by J. M. Skibo, W. Walker, and A. Nielson, University of Utah Press, Salt Lake City. ***Chapter 8:*** Studying Technological Differentiation, *American Anthropologist* (2002) 104: 1148-1161.

ISBN: 978-0-387-76524-2 (hardcover)
ISBN: 978-0-387-77132-8 (softcover)
DOI: 10.1007/978-0-387-76527-3

e-ISBN: 978-0-387-76527-3

Library of Congress Control Number: 2008920067

Cover illustration: "Colorful Cadenas," courtesy of Nathaniel Hardwick

Printed on acid-free paper

springer.com

Preface

The study of the human-made world, whether it is called artifacts, material culture, or technology, has burgeoned across the academy. Archaeologists have for centuries led the way, and today offer investigators myriad programs and conceptual frameworks for engaging the things, ordinary and extraordinary, of everyday life.

This book is an attempt by practitioners of one program – Behavioral Archaeology – to furnish between two covers some of our basic principles, heuristic tools, and illustrative case studies. Our greater purpose, however, is to engage the ideas of two competing programs – agency/practice and evolution – in hopes of initiating a dialog. We are convinced that there is enough overlap in goals, interests, and conceptions among these programs to warrant guarded optimism that a more encompassing, more coherent framework for studying the material world can result from a concerted effort to forge a higher-level synthesis. However, in engaging agency/practice and evolution in Chap. 2, we are not reticent to point out conflicts between Behavioral Archaeology and these programs.

This book will appeal to archaeologists and anthropologists as well as historians, sociologists, and philosophers of technology. Those who study science–technology–society interactions may also encounter useful ideas. Finally, this book is suitable for upper-division and graduate courses on anthropological theory, archaeological theory, and the study of technology.

The idea for this book came during a Fulbright sponsored trip to Porto Alegre, Brazil, by Jim Skibo in 2004, and conversations with Adriana Schmidt Dias, Fabíola Silva, Klaus Hilbert, and the participants in the seminar. The seminar was on ceramic analysis, ethnoarchaeology, and pottery use-alteration, but much of the discussion focused on how our theoretical approach to the study of technology contrasts with agency/practice, evolution, and other theoretical models currently in vogue. This led to a discussion between us and the decision to write this book. Although we had been collaborators for over two decades, this was our first co-authored book and we would like to thank the participants of the Fulbright seminar for the provocative conversation that convinced us to embark on this enjoyable enterprise.

A number of people commented on the manuscript: Eric Drake, Nathan Hardwick, Vincent LaMotta, Fernanda Neubauer, Charles Orser, Tim Pauketat, Michael Schaefer, and William Walker. Nathan Hardwick also created the wonderful cover art, "Colorful Cadenas." We also thank the editors at Springer, Teresa Krauss and Katie Chabalko, for their interest in this project. Finally, we thank our wives, Becky and Annette, for their unfailing support and love. This book is dedicated to them.

Normal, IL
Tucson, AZ

James M. Skibo
Michael Brian Schiffer

Contents

List of Figures

List of Tables

Chapter 1
People and Things: A Performance-Based Theory

A Kalinga man sat down next to the senior author during our evening meal. He had just returned from a day in his rice fields, repairing a terrace or performing some other dry season task, and I had spent the day with his wife and mother-in-law in their house watching them use pottery and inventorying their 20 or so vessels. Some of the pots, now two generations old and family heirlooms, had to be pulled down from the rafters and dusted off. We took each pot in turn, some obtained just 2 weeks ago and others up to 75 years old, and the women discussed in detail where the pot was made, who made it, and how it was used. Each pot had a story. Some of the pots were used to cook food at a recent funeral or wedding while others were used daily or even at every meal. The pots were originally obtained by barter for a few cups of rice, received as gifts, or inherited. Although the women were not potters, they could describe how the vessels were made and certainly comment on the quality of manufacture, the beauty of the design, and the skill of the potter.

There was no time to put away the pots before dinner, so they were placed on the bamboo floor in neat rows filling almost one quarter of the small, one-room house. As we feasted on rice and the freshly butchered chicken, the man looked with new interest at the vessels and, knowing my fascination with this technology, started to talk about the pots lined up against the wall. He told me things like how the vessels were made and where they were obtained, but with each of his statements, his wife and mother-in-law looked at me in dismay and shook their heads to indicate that he did not know what he was talking about. Finally, they could take no more as they laughed aloud and offered him a double-barreled wife/mother-in-law rebuke, the power of which seems to be universally understood. Sheepishly he looked down at his plate of rice but later leaned close to me and said in a low voice, "You should come up and see my rice fields." I took this to mean that I should abandon this silly work with pottery and look at a really important technology.

As archaeologists, we can empathize with that Kalinga man. If he did not know the technical choices involved in the manufacture and use of pottery, containers used around him everyday, how can we expect archaeologists to understand similar choices behind pottery, or any artifact that may be 2,000 years old from a behavioral system very unlike our own? The short answer is that it is not easy and we should not expect simple answers to any question about technology. The manufacture, use, and disposal of any technology – past or present, simple or complex – is woven into a social, economic, and ideological tapestry that is, in many ways, unique to a particular place and time. An investigation that reveals the tapestry's woven design, which in this case means the relationships between specific people and things, though not easy, is worth the effort.

In this book, we set forth a theoretical framework and offer case studies that serve as a guide in addressing the relationships between people and things. More specifically, the theory permits an understanding of the choices people make in inventing, developing, replicating, adopting, and using their technologies. A wide arc of factors, from utilitarian to social or religious, can affect these choices. Thus, our theoretical model provides the means to understand how people, past or present, negotiate these myriad factors throughout the life history of their material goods (for which we use terms such as artifacts, technology, and material culture more or less synonymously).

In this chapter, we furnish a general outline of the theoretical model's components, over 20 years in the making, including life history/behavioral chain, activities and interactions, technical choices, and performance characteristics. This is followed by a discussion, in Chap. 2, of how the theoretical model resembles but also differs from other archaeological models for understanding technological behavior, i.e., technological variability and change. Particular attention is paid to the "French school," selectionism, and approaches that employ agency and practice.

The best way to understand any theory is to observe it in practice, so the remaining six chapters are case studies that employ the theory (technically, we offer a "theoretical model" but will often just refer to it as a "model"). Chapters 3 and 4 focus on the origin, manufacture, and use of Native American technologies from very different parts of the continent, the American southwest and the Upper Great Lakes. Chapter 3 explores the origins of pottery on the Colorado Plateau between about AD 200 and 600. Pottery from this period is dominated by a globular, neckless jar often referred to as a "seed jar." By investigating the technical (formal) properties of the vessels (e.g., shape, size, temper, and surface finish), we establish the empirical foundation for understanding how the vessel could have performed cooking, storage, and transport functions. We conclude that the neckless jars were the Swiss Army Knives of the pottery world – a container that was used to carry out a variety of functions. Such a multifunctional vessel was well designed for the first-century settlement and subsistence system of mobile hunter-gatherers (see also Arnold 1999).

A use-alteration analysis (Skibo 1992) of the vessels themselves, focusing on carbonization and attrition, found that they had indeed been used in diverse activities. Sooting and internal carbonization patterns revealed that many pots were used to cook food both in wet (boiling) and dry modes (roasting or reheating food). Many vessels also had interior attrition consistent with fermentation (see Arthur 2002, 2003). The major point of this study is that, by combining use-alteration traces with a performance-based technical analysis, we were able to infer what the pots were designed for and how they were actually used.

In Chap. 4, we turn to the analysis of small pit features found during an excavation of *Gete Odena*, a small village site on Lake Superior occupied from 1000 BC to and beyond Euro-American settlement of the area. Similar pits have achieved notoriety in archaeological history, as Binford (1967) used them to illustrate his discussion of analogical reasoning in archaeological inference. Our analysis, however, focused on identifying the technical properties of the features and then grouping these properties into a set of performance characteristics. Based on this information, combined with other contextual clues (e.g., faunal remains, ethnohistoric documentation), we were able to show that the pits had been used to create a smudge fire for smoking and coloring hides.

In Chap. 5, we explore one aspect of technological change, the invention process in complex technological systems. The chapter proposes the "cascade" model of invention processes, which is illustrated by means of the electromagnetic telegraph, but it is applicable to any complex technological system, past or present. In the cascade model, performance problems in the development of a technology stimulate "spurts of invention." These invention processes continue to occur as new problems are encountered and resolved, which result in still more invention cascades. Although the model was first developed to handle industrial technologies, it can also be applied to traditional technologies. For example, invention cascades seem to have played a role in the development of ancestral Pueblo pottery after the initial seed jar form was introduced early in the first century AD.

Chapters 6 and 7 illustrate clearly that all activity links on the behavioral chain, not just those involving utilitarian or what we have called "techno-functions," can influence design and use. Regrettably, some researchers (e.g., Gosselain 1998:80–81) suggest that the model is focused only on utilitarian factors at the expense of social or religious ones, even though numerous studies have used the model to explore how topics such as gender (Chap. 7; Schiffer et al. 1994a; Skibo and Schiffer 1995), social power (Nielsen 1995; Walker and Schiffer 2006), religion (Chap. 6; Walker 1995a, 1998; Walker and Lucero 2000; Walker and Skibo 2002), and social class (Schiffer et al. 1994a) influence the design and use of material culture. This issue is taken up later in Chap. 2, but the point of Chaps. 6 and 7 is to illustrate how religious and social factors influence the design and use of technology in prehistoric and historic contexts, respectively. Chapter 6 investigates ball courts and ritual performance during the thirteenth century in southwestern New Mexico. A simple ball court feature was excavated, mapped, and analyzed using the performance model. The technical features of the ball court (e.g., size, location, orientation) are explained in terms of ritual performance. The ball court not only served to integrate the community in a village-wide ceremony, but it also placed the people within a religious interaction sphere centered at the large community of Casas Grandes located 60 miles south.

In Chap. 7, the rise and fall of the electric car around the turn of the twentieth century is explored with the model. By investigating a well-known transition, at least among historians of technology, this chapter illustrates that the theory can be used to create historical narratives and to reveal new insights into the causes of this technological trajectory. It is found that the performance characteristics of the technology, in relation to groups defined on the basis of social class and gender, ultimately led to the electric car's failure to penetrate a middle-class market.

The electric car had over a quarter of the market in 1900, but within two decades had almost completely disappeared as a commercial product. The electric- and gasoline-powered cars at the time are explored using a performance matrix, a framework for explicitly comparing any set of performance characteristics (e.g., utilitarian or symbolic) among competing technologies. The performance matrix clarifies which performance characteristics seem to have influenced acquisition decisions and also illuminates the resultant compromises in use. It is found that social class, wealth, and gender strongly affected the choices households made in their selection of gasoline- and electric-powered cars. The demise of the electric car was not, as some historians of technology contend, precipitated by the inevitable triumph of a "superior" technology (the gasoline-powered automobile), but rather is best thought of as a technological drama composed of many players who made choices about product acquisition based on their own activities and social position.

The final chapter demonstrates how the model can be used to understand large-scale patterns of technological change. Using eighteenth-century electrical technology as an example, this chapter offers a theoretical framework for studying technological differentiation, which is the process by which a new technology is transferred to communities, either between or within societies. The emphasis is on how people in these communities may redesign the technology to serve new social, symbolic, or utilitarian functions. Electrical technology was developed and expanded into a variety of communities in the eighteenth century, providing a good case study for developing this model, although it can be applied to any technology transfer, in simple or complex societies. A six-phase process of technology transfer is provided to serve as a guide to those wishing to apply the model to new cases of large-scale technological change.

Taken together, these case studies illustrate how the model can be employed to understand the relationships between people and things from the tinkering potter trying to create a vessel that serves as both a cooking and storage pot, to entire communities adopting and then redesigning a technology to achieve a new set of performance requirements. The chapters demonstrate how a performance-based model can integrate utilitarian, social, and symbolic factors important in the design and use of any material culture, simple or complex, in any society.

Behavioral Archaeology

Hegmon (2003) noted in a recent review of North American archaeological theory that there are few card-carrying members of "Behavioral Archaeology." This is, of course, true but she also notes that "many of the methodological and some of the theoretical insights of Behavioral Archaeology have been widely incorporated into various archaeological approaches" (Hegmon 2003:215). Although the goal of the original behavioral archaeologists, J. Jefferson Reid, Michael Schiffer, and William Rathje (Reid et al. 1974, 1975), may have been to conquer the entire discipline much the way new archaeology had done from the mid-1960s until the mid-1970s, this outcome was, in retrospect, highly unlikely. The discipline was undergoing a dramatic demographic shift; quite suddenly there were hundreds of new students aspiring to become professional archaeologists, all exposed to, and adopting in whole or in part, diverse theoretical perspectives (see O'Brien et al. 2005 for a detailed discussion of this period). Behavioral Archaeology came along at the beginning of an era that continues to this day, characterized by many different theoretical perspectives with currently little or no tendency toward unification (see also VanPool and VanPool 2003a:1–2).

The intent here is not to take the reader down memory lane but rather to make two points. The first is that by the 1980s, it was evident that for the foreseeable future Behavioral Archaeology would be but one of the several theoretical players in the dynamic field of archaeological method and theory. Behavioral archaeologists, in fact, seek to articulate with practitioners who approach the past from differing theoretical perspectives. The objective of *Expanding Archaeology* (Skibo et al. 1995), for example, was to demonstrate how Behavioral Archaeology can engage contemporary debates in the discipline. Indeed, we can also point to a growing body of researchers who have fruitfully employed various principles and heuristic tools from Behavioral Archaeology (see Hegmon 2003) or who have at least begun to think more deeply about the relationships between people and things (e.g., Arthur 2002, 2007; Beck and Hill 2004, 2007; Deal 1998; Longacre 1985; Longacre et al. 1999; Nelson 1991; Shimada and Wagner 2007; Silva 2008; Smith 2007; Varien and Mills 1997; Zedeño 1997, 2000). In short, the practice of Behavioral Archaeology, in this broad sense, is not predicated on archaeologists swearing allegiance to any particular theoretical program.

We are entering an exciting period in archaeological method and theory. As VanPool and VanPool (2003a:1–2) note, archaeology is at the point of shedding the Kuhnian legacy that suggested that our discipline's development would be characterized by dramatic paradigm shifts. Such a perspective created a great deal of rancor as champions of each approach attempted to denigrate the competition in order to score debating points. But after several decades of debate, a number of competing theoretical positions are as strong as ever and show no signs of imminent demise (O'Brien et al. 2005). We have reason to think that a similar theoretical environment will characterize archaeology for years to come. But this is not a bad thing. As O'Brien et al. (2005:253) note, "disparity of viewpoints in archaeology is not only refreshing but mandatory for the continued development and improvement of the discipline."

The second point, more relevant to this book, is that one need not become a behavioral archaeologist in order to employ the model we propose or to borrow any of its parts. Our model, we would argue, minimally can guide an analysis "for rendering the unknown knowable" (Walker et al. 1995:8) and serve as a starting point for crafting historical narratives regardless of one's theoretical orientation. As noted in Chap. 5, after the archaeologist conducts an analysis using the performance model, there is still "ample room for archaeologists who prefer, for example, agency, constructionist, or evolutionary explanations to craft their own narratives." Nonetheless, the foundation of this model is Behavioral Archaeology and it is important to outline, simply for heuristic reasons, the relevant elements of this theoretical approach (for other treatments of Behavioral Archaeology and uses of the model see LaMotta and Schiffer 2001; Reid et al. 1975; Schiffer 1976, 1995a, 2000, 2001a, 2008; Schiffer and Skibo 1987, 1997; Skibo and Schiffer 2001; Walker et al. 1995).

The core component of this approach is the redefinition of archaeology as a discipline that studies relationships between people and things in all times and all places. Pursuits such as ethnoarchaeology, experimental archaeology, the study of contemporary trash, or nineteenth-century technological change can be more easily brought under the umbrella of archaeology if we use this inclusive definition. Those who define archaeology narrowly as, for example, the study of prehistoric remains are excluding vital pursuits such as ethnoarchaeology as well as historical archaeology and denying the importance of an archaeological perspective for studying modern life (e.g., Rathje and Murphy 1992).

The focus for the behavioral archaeologist is behavior, the activities of everyday life. Behavior, however, is not conceived as merely the bodily movements of the organism, as in biology or psychology (Schiffer and Miller 1999a:11; Walker et al. 1995), but includes any artifacts participating in the interaction. "The animators of the behaving organism – culture, environment, and mind – take a backseat in behavioral archaeology to people's making, using, and depositing things" (Walker et al. 1995:5). It is these multifaceted relationships between people and things, one could argue, that makes humans distinctive (Schiffer and Miller 1999a:1–5). Beavers build dams, birds build nests, and chimps use and even make an occasional tool, but we are the only species that, figuratively, bathes constantly in an environment of our own artifacts. There are certainly many ways to investigate humans, past and present, but we argue that a behavioral approach has utility because "behavior – when defined to include both people and objects – ... mediates all ecological, social, and cognitive processes; through behavior the *potential* impact of extrabehavioral phenomena on life processes is *manifest*" (LaMotta and Schiffer 2001:20, emphasis in original).

The relationships between people and artifacts are discussed in terms of regularities discerned in processes of manufacture, use, and disposal that make up the life histories of material things, as in flow models and behavioral chains (Schiffer 1976). These relationships, sometimes described by law-like principles, have been criticized for being "essentialist" (e.g., O'Brien and Lyman 2000; Pauketat 2001). This criticism, however, seems to overlook the oft-stated claim that boundary conditions and specific behavioral contexts govern the applicability of specific principles (LaMotta and Schiffer 2001:24–27) and that these principles provide much of the content for what Wylie (1995) has referred to as the "source-side" of behavioral inference.

Another important component of Behavioral Archaeology is our contribution to understanding the cultural and noncultural processes that create the archaeological record (e.g., Schiffer 1983, 1985, 1987; LaMotta 1999; LaMotta and Schiffer 2005; Walker 1995a). The study of formation processes, probably the most widely used and appreciated component of Behavioral Archaeology, is based on the simple premise that behavior (not culture or mental states) proximately forms the archaeological record through making, using, and disposing of material items (Nielsen 1995). It is ironic that a failure to understand formation processes is often at the heart of the less-than-convincing case studies in contemporary archaeology. One fundamental flaw of the first case studies of the new archaeology, carried out on materials from Carter Ranch (Longacre 1970) and Broken K Pueblos (Hill 1970) in east-central Arizona, was that they did not explicitly consider how pottery was made, used, and disposed of in small pueblo villages, nor how such behavioral processes might have affected ceramic design-element distributions in the archaeological record (Schiffer 1989; Skibo, Schiffer, and Kowalski 1989). The irony is that archaeologists still make some of the same errors decades after we should have learned these basic lessons. Although we applaud attempts to explain archaeological variability and change in terms of social processes involving factors, such as gender, agency, and power, many of these arguments using prehistoric or even historic data are unconvincing because of the failure to take into account formation processes.

This discussion of Behavioral Archaeology, though brief, highlights its essential components. Archaeologists often borrow theories from other disciplines, such as cultural anthropology, sociology, and history, but it has long been the goal of behavioralists to construct their own theories and laws (e.g., Schiffer 1975a). Because we, as archaeologists, privilege the interactions between people and things and have access to data spanning millennia of human and technological change, we are uniquely positioned to develop theories for the invention, design, replication, and adoption of artifacts, i.e., the processes contributing to technological change (Schiffer 2004). Below, we offer one such theory.

The Theory

If the old Kalinga man did not understand his own pottery technology, how is it possible for an archaeologist, regardless of theoretical orientation, to pick up a pot or any object and begin to unravel the choices that people made in its design and use? This is indeed a difficult task but we take it even one step further. There are choices embedded in Kalinga pottery technology that even the potter cannot articulate, except to say something like, "This is the way we do it." We propose that even these choices, the reasons for which are now buried in the technological tradition, can be determined using the performance model. We realize, of course, that certain levels of understanding are more difficult to reach, especially when one is dealing with technologies lacking important contextual clues. Certainly, it is easier to unravel the choices in historical or modern situations when there is documentary information and other relevant evidence, but a lack of such information does not preclude the inference of decision making in ancient technology. This is because the first step in investigating a technology, past or present, is the object itself, which retains traces of manufacture – thus design – and use, which implicates adoption patterns.

In the chapters that follow, the reader will notice that we first discuss a technology's formal properties. For example, in investigating the earliest pottery on the Colorado

Plateau (Chap. 3), we begin with great details about temper, shape, thickness, etc., before attempting to infer why such choices were made. Similarly, in the ball court study (Chap. 6), the first task is to ascertain formal properties such as shape, size, orientation, and wall height before moving on to discuss how the technical choices are related to religion and ritual of the thirteenth-century Southwestern USA. Investigating technology transfer between electrical techno-communities in the eighteenth century is no different (Chap. 8). The investigator must discern the formal differences among examples of the technology and their functional variation as a foundation for understanding how and why it was differentiated when transferred to other communities.

Investigators who study historically known material culture enjoy obvious advantages when trying to understand technical choices. One could even argue that historians of technology are in a position to completely capture all the social, technological, and religious factors that impinge on the choices involved in the invention, adoption, and use of modern technology. If such were the case, however, we would quietly confine our model to traditional technology. But Schiffer and others have shown that this model can be applied with benefit to modern – even industrial – technologies, as is done in Chaps. 5, 7, and 8 (see also Gould 2001).

Because inferences about ancient technology often do not stand on equally firm footing as those about modern technology, we have spent a good deal of time exploring this inferential leap and trying to understand what we have termed "behavioral significance" (Schiffer and Skibo 1987). Miller's (1998:12–14) concept of "mattering" has some overlap with behavioral significance, and he demonstrates that even in modern ethnographic contexts it is up to the investigator to explore the significance of material culture in a particular setting, that is, how and why it matters. We concur. As archaeologists and students of material culture, we are always seeking – especially in ethnoarchaeology, experimental archaeology, and the study of modern material culture – to strengthen inferential links by understanding the behavioral significance of the material culture for the people who made and used it.

Our model helps one to investigate material culture variability and change. This is a fundamental consideration that underlies much archaeology and any discipline interested in the relationship between people and things. As discussed earlier, it is inconceivable to us that one can study humans without considering the material goods that surround them and, in fact, make them who they are. We define artifact very broadly as any human-made phenomenon from the stone tool and kitchen sink, to landscapes, televisions, and airplanes. Humans live in a material world of their making. Thus, this theory is applicable on any scale from the archaeologist trying to understand variability in Mousterian scrapers, Puebloan house style, and logging camp layout, to the engineer or social scientist interested in twentieth-century automobile design, the development of electrical systems, or changes in the space shuttle. This theoretical framework is designed to help us explain the technical variability of artifacts from the simplest to the most complex; it is applicable to artifacts that are both made and used by the same individual as well as those whose designers and manufacturers are far removed from the eventual user.

Our objective here is to provide a user-friendly version of the theory of artifact design followed by pertinent case studies. Interested readers should consult several other sources that explore other relevant aspects of technological change such as adoption processes (Schiffer 1992, 1996, 2001a, 2005a; Schiffer and Skibo 1987, 1997; Schiffer and Miller 1999a, 1999b; Schiffer et al. 2001; Skibo and Schiffer 2001; Skibo, Schiffer and Reid 1989, 1995). The design theory can be broken down into four components: life history/behavioral chain, activities and interactions, technical choices, and performance characteristics and compromises.

Life History/Behavioral Chain

The concept of life history is known in a variety of fields including engineering (Hughes 1990; Kingery 1993) in which product design models are broken down into the major steps, such as procurement, manufacture, and use, to identify performance requirements for a technology's various activities. In cultural anthropology, Kopytoff (1986) and others (Appadurai 1986; Thomas 1989; Weiner 1985) have introduced the "biographical" model of objects in which an artifact's exchangeability, importance, and meaning can change at various stages of life history (Orser 1996:107–129). In archaeology, lithic studies have long employed various life history models to describe the sequence of activities from raw material to finished form (see Bleed 2001a for a review). The life history approaches found in archaeology, engineering, and anthropology, however, are sufficient for only the most general explanations (Schiffer 2004:580). To understand a technology more completely, one must focus on specific activities and their constituent interactions, throughout its life, which we refer to as the "behavioral chain."

Behavioral chain has a long history in our work (Schiffer 1975b, 1976:49–53) and has been used in numerous studies in an attempt to isolate all links along an object's life history, each of which could have been important in its design (Schiffer and Skibo 1997). At this level, our concept of behavioral chain has much in common with the French concept of "*chaîne opératoire*" (Dobres 2000:154; Stark 1998:6), which was first introduced in English by Lemonnier (1986, 1992, 2002a). Acknowledging Cresswell (1976:6), Lemonnier (1986:149) defines *chaîne opératoire* as "a series of operations which brings primary material from its natural state to a fabricated state." This approach, traced back to the prehistorian Leroi-Gourhan (1943, 1945) and Mauss (1968), represents a long-standing and commendable interest in technology among a group of primarily French cultural anthropologists. As van der Leeuw (2002:240) notes, however, Leroi-Gourhan never published a definition of the concept of *chaîne opératoire*, possibly because it was simply understood as the operational sequence involved in the manufacturing process. Nonetheless, the way that it has been applied to understanding the relationships between people and things has much in common with our approach. We could not agree more with the statement by Mahias (2002:177), "Only by studying concrete operational sequences in a way that preserves these different aspects of technical facts will it be possible to uncover their underlying logics and to grasp in all subtlety just what makes them what they are."

Importantly, all links in a behavioral chain are considered as potentially relevant to a technology's design. Unlike *chaîne opératoire*, behavioral chains are not restricted to manufacture processes, for understanding design also requires attention to interactions in use, maintenance, reuse, deposition, and other postmanufacture processes. Chapter 3, which considers the origin of pottery on the Colorado Plateau, illustrates how different activities in the behavioral chain of a pot, such as storage, cooking, and fermentation, can influence that pot's design. A multifunctional vessel was designed as the potter peered down the behavioral chain and selected which performance characteristics to highlight in the design. In this case, we argue that the choices made were strictly utilitarian, but the potters could have highlighted any number of down-the-line interaction links. Vitelli (1999), for example, finds no evidence of cooking in some of the earliest Neolithic pottery and infers that the pots performed a nonutilitarian function. By investigating the design of the vessel and various contextual clues related to its behavioral chain, she argues that the pots had a ritual function and were likely made by shaman. Further discussion of differences and similarities between *chaîne opératoire* and behavioral chain are found in Chap. 2.

Activities and Interactions

The interaction between the Neolithic shaman and a pot, or the person wanting to first boil beans and then ferment beer in the same vessel, calls attention to the many kinds of links on a behavioral chain that can influence design. Each link is composed of an activity, which consists of specific interactions between people and artifacts, people and people, and even between artifacts. Schiffer and Miller (1999a:13), in constructing a behavioral theory of communication, add the concept of "externs" to potential interactions, which are various environmental phenomena that can play a role in an activity interaction. For example, the narrative constructed for the rise and fall of the electric car (Chap. 7) privileges specific behavioral interactions between people from various social classes, the electric- and gasoline-powered cars, and various externs such as rain that create muddy rural roads. It is these people, artifact, and extern interactions that help us to understand design differences between the two automobile types, which contributed to the demise of the electric car and the subsequent domination of internal combustion vehicles.

The set of social groups interacting with an artifact along its entire behavioral chain is called the "*cadena*" (Schiffer 2007; Walker and Schiffer 2006). *Cadenas* can vary in size from one person interacting with a few artifacts to hundreds or even thousands of artifacts and people. A behavioral chain analysis consists of identifying activities and then isolating the components of specific interactions, such as the type of people participating (social group), the location, frequency of performance, and other artifacts and externs (e.g., see Schiffer 1976, Fig. 4.4). Issues related to technological invention, design, replication, and adoption can only be resolved if one undertakes analysis at this level of specificity. The decisions artisans make in response to anticipated interactions along the behavioral chain are referred to as "technical choices."

Technical Choices

Individuals or groups have decisions to make when designing a pot, a ball court, or in altering a technology as it is adopted by their community. These technical choices lead to an object's formal properties and contribute to performance characteristics – that is, behavioral capabilities – in manufacturing and postmanufacturing activities. The choices can affect utilitarian performance characteristics, as when potters choose to add large amounts of temper to cooking pots to enhance thermal shock resistance, reducing the chance that the pot will crack over a fire. Technical choices also affect symbolic performance characteristics, as when a silversmith decides which shaping tools to use for producing either a cross or a Star of David. In Chap. 6, we discuss how the people at the Joyce Well site during the thirteenth century made choices in the construction of the ball court, such as constructing it away from the pueblo and orienting it to true north, reflecting both social and religious factors.

Lemonnier (1992, 2002a,b,c) and others (Bédoucha 2002; Cresswell 2002; Quilici-Pacaud 2002; van der Leeuw 2002) also highlight the concept of what they call "technological choice." Although these authors consider technical choices to be influenced by a wide arc of social, symbolic, and utilitarian considerations, the term technological choice conflates many processes, including design and adoption. Technical choice is a narrower and, for us, a more useful concept that pertains only to design processes – that is, the choices pertaining to procurement of materials and manufacturing activities. The case studies in Lemonnier (2002b) are often concerned with adoption processes. If technological choice referred only to adoption – people choosing among manufactured artifacts – then it would be a useful concept, though redundant with adoption, acquisition, or consumption.

Lemonnier (2002a) also highlights so-called illogical choices unrelated to practical concerns. The primary goal of these studies is to counter the notion that humans work like practical engineers solving problems in a "logical" way so as to emphasize that any technology is "first and foremost a social production" (Lemonnier 2002a:3). We certainly agree with that statement, having made it long ago (e.g., McGuire and Schiffer 1983), yet we cannot agree that choices, utilitarian or symbolic, follow *only* a culture-specific logic resistant to more general constraints and understandings. Our model, we would argue, furnishes a framework to understand both utilitarian choices as well the "nontechnical determinations" (Lemonnier 2002a:24), for all are embedded in a social and symbolic system. The Lemonnier model highlights the behavioral chain and the technological choices, as we do, but lacks additional concepts that permit a fuller understanding of technical choices as we define them. Additional differences between technological choice and our technical choice are discussed in Chap. 2, but the most notable difference, in terms of the present discussion, is that Lemonnier's model lacks the important concept of "performance characteristics."

Performance Characteristics

At each link in the behavioral chain – an activity – there are performances that facilitate specific interactions between people, artifacts, and externs. The focus is on the specific activities involved in the manufacture, use, and disposal of material culture. In this approach, the dichotomy between "function" and "style" becomes obsolete (Schiffer and Skibo 1997; see also Dietler and Herbich 1996:260; Lemonnier 1992:7; 2002a:10; Sackett 1977) as the material object itself is just one element that defines how it functions or what it "means." As Kingery (2001:130) notes, "A gold crucifix in the window of a jewelry shop plays a different role than the same object in the hand of a priest participating in blessing a condemned man facing electrocution." The social, symbolic, and utilitarian functions of an object are defined by its performances in activities all along its behavioral chain. For example, Pfaffenberger (2001) argues that symbolism is a consequence of technological activities, what we would call "performances." He demonstrates this point using ethnographic cases including yam storehouse use in the Trobriand Islands, whose cultural meanings are found in the artifacts themselves and also in the activities that produced them. Similarly, Dobres (2001), in a study of Paleolithic art, notes that the meaning of objects can only be known through the investigations of actions (see also Dobres 2000:134). What defines a ritual technology, according to Walker (2001), is not just its formal properties but also its performances in use and disposal (see also LaMotta and Schiffer 2001:42–44 for a discussion of performance in disposal). For example, an everyday cooking pot may boil beans in the morning and be used in a shamanic ritual at night, or be placed on the floor as an offering just before a house is burned (Walker 1998, 2001; Walker and Lucero 2000; Walker and Skibo 2002). The function of this pot - and what its visual performance may communicate - changes as it moves along the behavioral chain. A complete understanding must consider, to the fullest extent possible, how an object performs during all links in the behavioral chain.

In order for an object to perform in an activity at an acceptable level it must possess certain capabilities that we have termed "performance characteristics" (Schiffer and Miller 1999a; Schiffer and Skibo 1987, 1997; Skibo and Schiffer 2001; Skibo, Schiffer, and Reid 1989). This term was first introduced into archaeology by Braun (1983), who was referring to utilitarian performance alone (for a recent illustration of utilitarian performance see Pierce 2005). Competent utilitarian performance is necessary for an object's successful interaction in a specific activity. For example, a cooking pot must have high thermal shock resistance to withstand repeated placement over an open fire without cracking. Thermal shock, in this case, would be a primary and heavily weighted performance characteristic. To increase the likelihood that the pot will not crack over a fire, the potter – as a designer – would have at his or her disposal a number of possible technical choices such as adding more temper and making thinner walls (Pierce 2005; Skibo, Schiffer, and Reid 1989). But utilitarian performance characteristics such as these are just part of what goes into the choices that people make in designing, using, and acquiring a technology. There is also a large set of "sensory" performance characteristics relating to sight, touch, smell, hearing, and taste that can be important in an activity and thus have an impact on design and use.

Visual performance characteristics are rather common in the interaction between people and things and influence technical choices. Kalinga cooking pots from the Pasil Valley have a rounded shoulder in contrast to those made in the nearby Tanudun Valley where pots have an angular shoulder (Longacre 1991; see Stark and Skibo 2007 for a review of Kalinga ethnoarchaeology). If a novice potter from the Pasil Valley started to make a cooking pot with an angular shoulder, she would be corrected; and even if the vessel were fired it would be rejected by Kalinga consumers. A successful Kalinga cooking pot must be thin walled and coated on the interior in order to achieve a number of cooking-related performance characteristics, but it also must have a certain profile, height-to-width ratio, and incised design that communicates to consumers that this vessel was made in the Pasil Valley by a Kalinga potter. Some of these visual performance characteristics are important down the behavioral chain as the pots in use, stacked on the shelf or carried from the water source, communicate to any observer that this person uses pots made in the Pasil Valley by a Kalinga potter.

In the Schiffer and Miller (1999a) model, one can isolate different types of communication between people using Kalinga pots during different activities on the behavioral chain. When a potter comes to a nonpotting village to sell pots (see Aronson et al. 1994; Stark 1994), one can identify three interactor roles (sender, emitter, and receiver). Visual performance characteristics play a role in pottery selection in terms of both utilitarian and social factors. The sender is the potter carrying to the village a collection of pots (emitter) that is involved in a communication process with the potential pottery consumers (receiver). The buyer can identify a Pasil pot immediately based on various visual performance characteristics imparted by the potter's technical choices. If the buyer wants a vegetable/meat cooking pot, these too can be identified by the vessel's visual performance; a vegetable/meat pot is squatter and has a more open orifice, which both permits easy access to the contents and also reduces heating effectiveness slightly so that water simmers without boiling over.

Once the pots are purchased, interactors in the communication process change slightly. The woman, carrying these pots to her rice fields or to another village for a funeral, silently communicates to people she meets on the trail that she is from the Pasil River Valley, and this is not a trivial matter. These seemingly simple communications can have life or death consequences during times of tribal war when raiders hide along trails to ambush someone from a particular region as part of the ongoing blood feud (see also Wobst 1977). Although this example simplifies the communication process (the people involved in this interaction have at their disposal many other verbal and nonverbal means to communicate), it demonstrates how the communication processes involving people and artifacts vary as one travels down the behavioral chain.

In Chap. 6, we illustrate how social and religious messages are communicated through visual performance of a ball court's formal properties. The occupants of the thirteenth-century pueblo at Joyce Well were composed of different kin groups probably represented by the site's three plazas. Previously, people in this region of southwestern New Mexico lived in very small pueblos, yet in the thirteenth century there was a tendency to aggregate into larger communities, which inevitably led to new kinds of social tension. We believe that the ball court, which is not immediately adjacent to the pueblo and is thus not likely to have been aligned with any particular social group, may have functioned to ease these tensions by integrating diverse community members in the context of a ritual ball game. The visual performance characteristics of the ball court, when in use, may have facilitated the conveyance of information among players and spectators, to themselves and to each other, about their own membership in a single community entity. In a similar way, but in a modern context, Pellegram (1998) demonstrates that paper-use behavior in a London office not only conveys messages to the people in the office but also that the stratified organization of the office was communicated and reinforced.

The orientation (true north) and other formal properties of the court are also involved in visual performance. The community as a whole is communicating to all members and to any outsider that its members are involved in a ritual that links them to the Casas Grandes religious interaction sphere (Walker and Skibo 2002). In this case, the visual performance characteristics of the ball court are essential and heavily weighted in its design. To understand how these visual performance characteristics function in this society, it is important to infer the senders and receivers of this communication. Jarmon's (1998) investigation of banners and flags used in Northern Ireland parades provides a good example of how strong political and social messages are communicated through visual performance. To understand this communication process, he not only had to clearly delineate both the senders and receivers of these messages, but also employ a life history approach to fully appreciate the role of the banners and the social groups involved. The adoption of electric lighting in some nineteenth-century lighthouses provides a similar example (Schiffer 2005b). Electric arc lights were distinctive in color and brightness compared with oil-burning lamps, which made it possible for them to communicate meanings such as the nation's commitment to modernity, safe navigation, and the ability to manage cutting-edge electrical technology.

Sometimes, visual performance characteristics provide unintended negative responses from potential users, as receivers, that can result in very limited adoption of a new technology. The shirt-pocket portable radio of the 1940s is a clear example of such a failure (Schiffer 1991:168–169). This new form of portable listening fit into one's pocket and was listened to through a small earplug. Although similar devices are common today, this radio was rejected in the 1940s because of visual performance. The radio and associated earplug looked exactly like hearing aids of the day and the user of the device did not like what it was communicating to people on the street – that they were hard of hearing.

No design is perfect because there is a complex relationship among technical choices, technical properties, performance characteristics, and life history experiences. In terms of utilitarian performance, cooking pots demonstrate how some properties may positively influence one performance characteristic while at the same time deleteriously affect others. As noted earlier, because thermal shock resistance is such a heavily weighted performance characteristic in cooking vessels, potters make technical choices that increase the likelihood that the vessel will survive when placed repeatedly over a fire, such as increasing the amount and size of mineral temper and decreasing wall thickness. As one adds ever more temper and decreases thickness, there comes a point when the strength of the pot decreases beyond what is acceptable. Consequently, compromises must be made in design. Enough temper is added to give the pot adequate thermal shock resistance but not so much that it might break from even the slightest impact. In any design, it may be that no performance characteristic is ever achieved at an optimal level. In addition, different communities may want to stress certain performance characteristics at the expense of others or have different perceptions of adequate performance. Chapter 8 illustrates how this process can lead to changes in a technology as it is transferred from one community to another.

In cases where visual performance characteristics are heavily weighted, it might be acceptable to make significant compromises in utilitarian performance. The Kalinga in the late 1980s began to use metal pots for rice cooking as aluminum conducted heat far better than ceramic and thus could heat rice faster (Skibo 1994). More important, they did not use the pots for vegetable/meat cooking because the ceramic pots were better at simmering without boilover, whereas the aluminum pots could heat the water too quickly, boil over, and douse the fire. Visual performance comes into play here as the women insisted on removing all the exterior soot on the aluminum pots after each use. As any camper knows, soot actually impregnates the metal itself and, in removing the dark carbon layer by vigorous scrubbing, a thin layer of aluminum is also removed. Metal pots were a sign of wealth and modernization in Kalinga households, and so shiny aluminum pots were hung from the rafters or stored in conspicuous places for all to see. Maintaining a shiny metal pot, however, requires vigorous washing with sand as an abrasive. Not only does this take significantly more time but the pots are also worn through in a relatively short period. Visual performance in display activities, in this case, was heavily weighted at the expense of utilitarian concerns – vessel longevity and ease of cleaning – as the women were seeking to communicate with people entering their home that they were modern and had a certain amount of wealth.

Application

How does one ferret out these various performance characteristics, especially those that might be heavily weighted in social and ritual activities? The purpose of the following chapters is to demonstrate how the model is applied to understanding technological variability, design, and change in a variety of contexts. In each case, the concepts discussed earlier (behavioral chain, activities and interactions, technical choices, compromises, and performance characteristics) are used to move beyond vacuous concepts such as "style" or "function" to a more meaningful understanding of the relationships between people and things. Each case study goes about this process in a slightly different way, but all are based on the behavioral model. They all begin with the technology itself, and then move to understanding the technical choices of the producer in the social and environmental context in which it was made. Isolating these decisions requires that we peer down the behavioral chain and explore the activities and interactions involved in this technology.

In some studies, a final component that comes into play is that of summarizing technical choices and activities in terms of performance characteristics. This process can be fostered by creating a performance matrix (Schiffer 1995b, 2004, 2005b; Schiffer and Skibo 1987), which aids in making comparisons among competing technologies. The relationships between people and things are profitably conceived as a set of performances, occurring at micro and macroscales, played by artifacts and individuals or groups of people trying to make a technology that works at a utilitarian level, yet is made and used in a social context unique to a time and place. We should note, however, that the kind of analysis we propose is not easy at any level – especially with prehistoric data. In fact, the case studies in this volume that use prehistoric data (Chaps. 3, 4, and 6) should only be considered first attempts to understand the performance of a technology in its many behavioral and social dimensions. We anticipate that a more complete understanding of how and why globular pottery appears, ball courts were made, and smudge pits were used will accrue as researchers reorient their excavation and overall analytic strategies to collect the types of contextual information that are required for a rigorous performance-based analysis.

Chapter 2
Behavior, Selection, Agency, Practice, and Beyond

Chapter 1 outlined our theory for studying technological variability and change. In this chapter, we address other models that deal with these issues. Although there are a number of approaches to understanding the relationships between people and things (e.g., Broughton and O'Connell 1999; Fitzhugh 2001; O'Connell 1995), here we focus on three major schools of thought that are common in contemporary archaeology: Evolutionary Archaeology, what we call the "French school," and agency and practice theory. The relationship between selectionism, one variant of Evolutionary Archaeology, and Behavioral Archaeology has been explored previously (O'Brien et al. 1998; Schiffer 1996), so those arguments need to be reviewed but briefly here. We will, however, spend some time on the relationship between our model and the French school. Although we have already discussed some areas of overlap between these approaches, there is need for a more detailed discussion. Finally, we review how agency and practice theory have been applied to the understanding of material culture change. These approaches are quite compatible with our model and can be integrated in a useful way. This chapter concludes with several examples that illustrate a behavioral strategy for investigating social power.

Evolutionary Archaeology

As already noted, similarities and differences between Behavioral Archaeology and Evolutionary Archaeology (also referred to as selectionism) have been aired elsewhere (O'Brien et al. 1998; Schiffer 1996; see also O'Brien 2005 and O'Brien and Lyman 2000 for a recent summary of Evolutionary Archaeology), and others have criticized the selectionist framework from different perspectives (e.g., Arnold 1999a; Bamforth 2002; Boone and Smith 1998; Pauketat 2001; Spencer 1997; Wylie 1995, 2000). The most important area of overlap, in terms of our model, is the focus on artifacts and utilitarian performance-based interactions. Selectionists have also used a form of the life-history approach in tracing phenotypic features (O'Brien and Holland 1992:52; O'Brien and Lyman 2003b). These general life-history approaches, as we discussed earlier, are instructive but often are not specific enough to isolate discrete links in the behavioral chain that influence, for example, design and adoption processes.

J. Skibo and M.B. Schiffer, *People and Things: A Behavioral Approach to Material Culture*

Although there is neither one monolithic selectionist model nor just one behavioral model for explaining technological change, one can still identify two major incompatibilities that are relevant here. The first has to do with the nature of inference (O'Brien 2005:31; Schiffer 1996:650–652). Inference, as defined by Behavioral Archaeology, and reconstructions of the past have been deemed by selectionists as unscientific "just so" stories (Dunnell 1980, 1982, 1989; Neff and Larson 1997:77; O'Brien and Lyman 2000:346–348) because they typically rely heavily on functional principles developed through ethnoarchaeology and experimental archaeology. According to O'Brien and Holland (1995:178–179), "any search for universal laws that govern behavior is not only incompatible with an evolutionary approach but is doomed to fail." We discussed this earlier but we will note again that functional principles and various behavioral regularities are not typically universal but rather have specific boundary conditions defined by critical variables (LaMotta and Schiffer 2001:24–25). The more orthodox evolutionary archaeologists understand the notion of boundary conditions completely but are still unconvinced. According to O'Brien and Lyman (2000:347), behavioral reconstructions "may be real, or sort of real, or not at all real; we simply have no way of knowing because of the shaky ground….upon which they are constructed." They say further that, "universalities not only *do not* exist, they *cannot exist*" (O'Brien and Lyman 2000:349, emphasis in original). This is a hard-line perspective that undermines the inferential process in archaeology and even denigrates the works of some who carry the evolutionary archaeology banner (e.g., Graves and Ladefoged 1995; VanPool and VanPool 2003b).

It is true that behavioral inference is based on numerous principles – general and specific – many of which were developed in actualistic studies. For example, the foundation for the analysis of the earliest pottery on the Colorado Plateau (Chap. 3) is inferences about the intended and actual functions of the vessels. The *intended* function of the pot was inferred by isolating the technical choices based on various properties (e.g., temper size, type, and amount, and vessel shape and size). We concluded that the pot was designed to perform many utilitarian functions. This simple, yet important, inference is based on nomothetic principles developed through experimentation (e.g., Pierce 2005; Skibo et al. 1989b), ethnoarchaeology (e.g., Arnold 1985, 1993; Arnold 1991; Kramer 1982, 1997; Longacre 1991, 1999), and in ceramic ecology and ceramic science and engineering (Kingery 2001; for an overview see Rice 1987, 1996a,b). To infer the *actual* functions of the vessels, a use-alteration analysis was done, which was based on principles developed both in ethnoarchaeology and experimental archaeology (Skibo 1992; see also Arthur 2002, 2003, 2007).

A second incompatibility is selectionists' obvious uneasiness when having to deal with the nonutilitarian aspects of technology. The most convincing selectionist case studies deal with variability and change that is best explained in terms of utilitarian performance characteristics (e.g., Dunnell and Feathers 1991; Feathers 2006; O'Brien et al. 1994; Pierce 2005; Van Pool and Leonard 2002). VanPool and VanPool (2003b:107) note that, "An EA approach is ideal for explaining patterns and subsistence factors that have strong implications for replicative success, but it is likely to be less intuitively satisfying when seeking to explain the changes in ceramic decoration and the development of social hierarchy." Selectionists do consider decoration or style but only as part of the "nonadaptive aspects of phenotypic variation" (Neiman 1995:7) that then can be used to construct historical lineages based on Darwinian principles such as drift (Neiman 1995; O'Brien and Lyman 2000, 2003). According to O'Brien and Holland (1995:190), "engineering-design analysis offers an appropriate basis from which to construct plausible... arguments relative to fitness, and thus overcome the 'just so' hurdle." This is because in the cases they have selected, we would argue, the primary performance characteristics (i.e., those that are most heavily weighted in the performance matrix) are utilitarian. These studies focus on the utilitarian performance characteristics that influence an artifact's replicative success (using their jargon). Very similar studies have been done using our model (e.g., Skibo et al. 1989b), underscoring an area of overlap mentioned earlier.

Behavioral Archaeology and Evolutionary Archaeology, however, part ways when dealing with nonutilitarian performance characteristics and what Dunnell (1989) has called "waste," which is human behavior that is not directly tied to biological reproduction and replicative success. Our model obviously handles utilitarian performance characteristics very well but, more importantly in the present context, it also easily incorporates various social and ideological factors that may also affect the design and use of any technology.

A notable exception is the study by Graves and Ladefoged (1995) in which they apply evolutionary principles to the study of ceremonial architecture, which they prefer to call "superfluous" instead of "waste" behavior. They argue that ceremonial architecture, in this case, had an important functional role by reducing risk under conditions of resource stress. Similarly, VanPool and VanPool (2003b) combine an evolutionary approach with agency theory in their investigation of ritual technology associated with the northern Mexican site of Casas Grandes. Evolutionary Archaeology is used to explain changes in subsistence and settlement, whereas agency is used to understand symbolism and the development of social inequality. The authors recognize the limitations of Evolutionary Archaeology and seek to fill the gaps with other theories. This move, we note, brings them closer to a strategy advocated here.

In the words of Dunnell (1989:46), "Evolutionary theory is not, at present, capable of explaining much of the archaeological record." Nonetheless, the historical narratives offered by selectionists, which invoke natural selection and drift, are sometimes well developed and convincing. The fundamental problem with Evolutionary Archaeology may be that many other researchers are interested in aspects of the archaeological record that selectionists consider either uninteresting or unimportant. Although there have been some attempts to stretch evolutionary theory or to combine it with other theories, for the most part Evolutionary Archaeology retains a narrow focus (Sillar and Tite 2000:15), and thus many archaeologists find it unattractive.

The one trait that many archaeologists, including us, inherited from the New Archaeology of the 1960s was optimism (see also Chap. 4). The enduring legacy of that period is that much of the human past is potentially knowable; we just need to develop the method and theory for establishing rigorous inferences (Binford 1968). We have not lost the optimism, whereas the core of evolutionary archaeology seeks to restrict archaeological investigation to a limited range of subjects, with a correspondingly limited range of causal factors. Our goal is to develop archaeological method and theory, and to keep showing how our model can be employed to investigate all the factors involved in archaeological variation and change. We continue to have the optimism of New Archaeology that all is potentially knowable.

French School

There is not one monolithic school of thought in French archaeology regarding technology, but we use the phrase "French School" (see also Wilk 2001) as a shorthand to designate an approach that combines the use of *chaîne opératoire* with the "social production of techniques," as articulated by Lemonnier (1992, 2002a), and now integrates social agency and practice (Dobres 2000). One could argue that selectionists, and the science-based histories that they produce, are on one end of the continuum, while the French School, which tends to highlight social factors in explanations of artifact variability and change, and downplay utilitarian ones, lies at the other end. There is a good deal of overlap between the French School and Behavioral Archaeology, and much to admire in their case studies, but we would argue that the French School, nonetheless, has shortcomings that prevent it from developing a comprehensive understanding of the relationship between people and things. We turn to a comparison of the two approaches. In Chap. 1, we briefly mentioned some areas of overlap between the French School and our model. This included the concepts of *chaîne opératoire* and technological choices, both of which, in a slightly different form, play a key role in our model as well. We made the point, however, that because their model lacked the concept of performance characteristics (primary and secondary) and design compromises, their case studies often conclude that people make "illogical" choices in their technology. These differences, we argue, can be traced to a single issue: the concept of *chaîne opératoire* is too narrow. It lacks a historical perspective, and fails to take into account an artifact's life history beyond manufacture.

Lemonnier (2002a) begins the introduction to the volume *Technological Choices: Transformation of Material Culture since the Neolithic* with a quote from Quill (1985:7–8) that discusses the British tradition, in the early 1930s, of landing aircraft on a gliding approach with the engine throttled back. This technique, they believed, would train pilots to land planes in the event of an engine stall. The problem was that in this procedure, used by no other air force, pilots often lost control during landing and crashed, resulting in many deaths. In fact, more pilots died during everyday throttled-back landings than in emergency landings as a result of engine failure. Eventually, the British realized the error in this procedure and by the late 1930s they trained their pilots to use the safer "power-on" landings.

This extreme and "absurdly negative" (Lemonnier 2002a:1) example sets the tone for the entire book, which focuses on how "social logics unrelated to technology may weigh heavily on the evolution of technological systems" (Lemonnier 2002a:2). "Technology" in this case refers to what we would call utilitarian performance, and looking at this example from the strictly utilitarian perspective does indeed make the choices in landing procedure seem illogical. Social logic, according to this perspective, often trumps utilitarian logic. Because technology, in Lemonnier's (2002a:4) view, is "a complex phenomena in which wide symbolic considerations are involved from the start, it becomes tricky to separate the 'technical' from the 'social'." It is indeed "tricky" and we would add that it is even trickier with the narrowly construed *chaîne opératoire*, which derives from, and appears to be most useful for handling, ethnographic situations.

Social logic appears to trump utilitarian performance only when viewed from a present-day ethnographic perspective. In this and other cases, the *chaîne opératoire* lacks the historical perspective required to garner a more complete understanding of, for example, the manufacture and use of British aircraft during the 1930s. In our model, there is no such thing as "illogical choices." According to Lemonnier (1992:17), however, "it is as if, during its history, a society, for unknown reasons, had come to rely on one particular technique." Because behavioral chain is historical in nature, these "unknown reasons" are of great interest and potentially knowable. As Roux (2003:5) notes, the approach by Lemonnier and others "pursues a problematic scientific position by taking into account only certain kinds of evidence while ignoring others." She goes on to argue that, "different technical solutions met in the history of aviation may thus be interpreted in terms of arbitrary choices when decisive technical, economic, and environmental parameters are ignored" (Roux 2003:5). We agree with her on this point: if one focuses too heavily on the social and investigates a technology from the ethnographic present, then the researcher could be missing the underlying causes of a seemingly "illogical" choice. We would take this point a bit further by stating that even if the reason for a choice is rooted in the social, it does not mean that it is illogical or that it will defy our understanding even in the distant past. In the Kalinga metal pot example described in Chap. 1 (Skibo 1994), polishing one's pots is an illogical choice from the perspective of utilitarian performance – it takes longer to wash these vessels and eventually wears a hole in the pots. However, if one looks at it in terms of visual and symbolic performance in the context of display activities, the choice to shine the pots is quite logical. Framing an investigation in terms of logics seems to unnecessarily restrict the researcher. Consequently, we prefer the concept of "choices," which reflect various weightings of performance characteristics that can be related to any combination of social, religious, technological, and political factors.

From a historical perspective, choices build upon choices – all made in the context of people's traditional knowledge and social system (see also Sillar and Tite 2000:5). A technology that "works" (i.e., achieves acceptable levels of relevant performance characteristics) will continue to be replicated until someone or a group decides that it is no longer working at an acceptable level. Because we combine a focus on the artifact at a particular time and place with a historical understanding of the people and their activities, we can create a more well-rounded model of the choices made and the factors – of diverse kinds – that influence them.

A complete understanding of a particular technology must also incorporate primary and secondary performance characteristics and the important concept of compromise, all of which are lacking in the French model. No design is "perfect" because of the many diverse interactions in which it must perform during its life history. Tools do not operate at some peak level and, consequently, many of the technical choices seem to be illogical. Why pick a plane-landing procedure that kills people? Because in the short term it was operating at an acceptable level that included, unfortunately, the taking of human life. It was "logical" only in terms of initial assumptions made about the safety of the procedure and the dangers of dead-stick landings. At a certain point, new information had accumulated that led to a revision of assumptions and a change in the procedure.

A fundamental problem with *chaîne opératoire* is that it does not continue through use activities and beyond. In our design model that applies to the behavioral chain, one needs to explore postmanufacturing activities to understand technical (not technological) choices. Behavioral chain, unlike *chaîne opératoire*, also takes in formation processes. In archaeology it is essential to have a behavioral chain that begins with procurement of materials for manufacture but continues after use, through deposition, until recovery by a researcher (see Chapman and Gaydarska 2007). Thus, reuse processes and other post-initial use formation processes, including those of the natural environment, enter into a complete behavioral chain. According to Lemonnier (2002a:24), "we no longer have the means of digging up those non-technical determinations of techniques that nevertheless resulted in highly efficient artefacts or procedures." Privileging the social and cultural, however, provides an incomplete picture of technology (Roux 2003) in the same way that evolutionary archaeologists tend to focus on the narrow reality associated with utilitarian performance. *Chaîne opératoire* lacks the expanded life history approach that is essential to archaeology (Bleed 2001a). Our approach privileges nothing – it is causally agnostic – and so enables the researcher to consider *all* potentially relevant factors at the interface of people and things.

The concept of *chaîne opératoire* as originally defined by Lemonnier has been adapted by some to suit research interests. Dobres (2000:155) notes that if *chaîne opératoire* is to be useful in archaeology, meaning and sociality must be inserted "into descriptions of physical sequences of material transformations." She argues that *chaîne opératoire* "highlights the sequential nature of both material and social reproduction," thus bringing in the concept of social agency (Dobres 2000:156; see also Dietler and Herbich 1998; Dobres and Robb 2005:163).

Agency and Practice

Our performance-based approach, which focuses on choices that individuals or groups make in the design, manufacture, adoption, and use of a technology, has always highlighted human agency (Sillar and Tite 2000:9). Hodder and Hutson

(2003:33–35) acknowledge that the behavioral approach does use the concept of agency especially when exploring technological change, yet they state that we have gone "too far in that direction" (Hodder and Hutson 2003:34). We give our potters, according to Hodder and Hutson (2003:34), "unrestricted latitude for experimentation." This criticism, that we believe that people who make and use technology operate like practical engineers making tests and solving problems, is often mentioned (e.g., David and Kramer 2001:141; Gosselain 1998), but it is a caricature of the behavioral approach. If the reader has gotten this far in the book, it should be clear that we do not advocate such an approach. Our concepts of performance characteristic, technical choice, and compromise include the notion that people make decisions about their technology based on their knowledge, experiences, and the social and natural environment in which they live. And the experiments that people carry out are equally contingent upon local circumstances.

When our approach isolates connections between people and things, Hodder and Hutson (2003:33) dismiss them because these interactions "silently contain modern western assumptions about the meaning of the artifacts" (see also David and Kramer 2001:141). Well, of course they do. As anthropologists we understand that the concept of culture works both ways. We are embedded in our own culture equipped with biases and agendas that can influence all we do, including reconstructing technical choices made by prehistoric potters. Because of this very problem, we introduced the concept of "behavioral significance" when trying to apply results from our laboratory experiments to studies of prehistoric pottery (Schiffer and Skibo 1987; Skibo et al. 1989b). Through these experiments, for example, we could show that large amounts of sand temper would give a pot greater thermal shock resistance, which would mean that a pot could be placed over the fire many more times without failure. The problem, however, is determining whether these differences in expected use-life are "behaviorally significant." That is, would a pottery user be able to realize this difference in use-life, and, more importantly, would the potter add large amounts of sand temper to increase thermal shock resistance? Realizing that these types of relationships are created in our lab and the whole experiment is set up by us, products of our own culture, we seek to know if these differences were actually taken into account by the pot makers and users. The short answer is, we can never know with complete certainty. We can only make inferences, which are arguments based on relevant evidence and relevant principles. Others can critique these inferences, but to dismiss them because they may contain western assumptions is lazy scholarship. All of archaeology and, in fact, all of science rest upon equally contestable assumptions. Indeed, it could be argued that the belief that traditional societies depend on different logics and that social and ideological factors dominate technological decision-making reflects the western colonialist view that "the other" is qualitatively different from us.

A more productive strategy is followed by a number of researchers who are attempting to "do" agency and practice theory with archaeological data. No one has done more for advancing agency in archaeology and especially the relationship between technology and social agency than Marcia-Ann Dobres and her colleagues (Dobres 1995, 2000, 2001; Dobres and Hoffman 1994, 1999; Dobres and Robb

2000, 2005). Agency in archaeology, however, has been defined and used in so many different ways that it has perhaps rendered the term virtually useless (Dobres and Robb 2000:10; Dornan 2002; Pauketat 2001). Dobres and Robb (2000) advocate a more restricted definition of agency, and a good deal of their work has focused on defining it in ways that make it useful to archaeologists (Dobres and Robb 2005). Despite these efforts, there is still much variability in how one defines and uses this approach (Dornan 2002). For example, Dobres and Robb combine agency theory (*sensu* Giddens) and practice theory (*sensu* Bourdieu), while others believe that this is a distinction that should not be blurred (Orser 2004:126; Pauketat 2000:114).

Advocates of agency or practice theory have, on occasion, maintained that their approach is opposed to the behavioral model. Some have argued that a behavioral approach with its so-called "essentialist" underpinnings is at odds with practice theory (Pauketat 2001). Pauketat (2001:76–79) maintains that we should focus on what people do and how they do it rather than why they do it. Following Dobres and Hoffman (1994), he argues we need to free ourselves from "behavioral essentialism" and focus instead on technology as socially negotiated practices, which he believes is quite different from a "behavioral position" (Pauketat 2001:78). He also notes that behavior is "little more than....action with predetermined courses and predictable ends analogous to other times and places" (Pauketat 2000:115; see also Pauketat 2003:41–43). This view does not indicate an understanding of the concept of boundary conditions and fails to appreciate that behavioral principles, from general to specific, are the foundation for many archaeological inferences (see also Roux 2003, 2007). We appreciate the generality of nomothetic principles and general concepts, and at the same time acknowledge the importance of local, situational, and contingent factors. An appreciation for nomothetic principles is not incompatible with a concern for local contexts. These common yet erroneous caricatures of our concept of behavior have resulted in "behavior" becoming an epithet. According to Wobst (2000:40), at the "Theoretical Archaeology Group Meetings in Durham England, presenters engendered intense negative reactions if they unthinkingly let the term 'behavior' slip into their remarks." These types of caricatures get started in such meetings, are blindly repeated in graduate seminars and discussed in hallways and coffee shops, and then become facts without many taking the trouble to read the original works. Although it is always useful to highlight differences in theoretical approaches, we strive to move beyond this polemic, seeking to engage works that employ agency and practice theory in a search for points of convergence rather than divergence.

According to Pauketat (2000:115), practice theory "is a theory of the continuous and historically contingent enactments or embodiments of peoples' ethos, attitudes, agendas, and dispositions." In this context, practices or "negotiations" (Pauketat 2000:116) are quite similar to our concept of performances of people and artifacts along a behavioral chain, which is implicit in "the relationship between people and things," a phrase at the core of Behavioral Archaeology. Walker and Lucero (2000) are comfortable in merging a behavioral model with agency in their study of prehistoric ritual and power. Their strategy is to "highlight how agents organize material culture in pursuing various activities including raw materials acquisition, manufacture, use, and discard" (Walker and Lucero 2000:130). They see no problem in examining how agents create life histories of artifacts *nor do we* because such a perspective has always been part of our approach.

Pauketat (2001) and others also have distaste for the way we might isolate various goal-oriented behaviors. This seems to be a reference to our inferences about various technical choices such as potters designing a vessel for multiple functions, as is done in Chap. 3. In contrast, Pauketat (2001:79) suggests that we jettison such goal-oriented action and focus on practice, which is guided by, among other things, "doxic" referents (*sensu* Bourdieu 1977, 1990; Giddens 1979, 1984). Doxic referents are various forms of knowledge that include unconscious, spontaneous, nondiscursive, practical, and commonsensical. More than two decades ago, we identified (Schiffer and Skibo 1987:597–598) three essential components of technological knowledge, recipes for action, teaching frameworks, and techno-science, which are very much like the doxic referents. Recipes for action are the rules, tools, and sequence of actions that underlie the production of a technology. The teaching framework is the practices that permit the transmission of the technology intergenerationally, and techno-science is the principles that underlie an artifact's successful performance. Roux's (2003) concept of "technological fact" is also quite similar to doxic or technological knowledge. In Chap. 3, which describes the origins of pottery on the Colorado Plateau, we are purposefully focused on this type of technological knowledge. In earlier work (i.e., Skibo et al. 1989b), which included experimentation, we also investigated technological knowledge and the technical choices of cooking pots because we found the current explanations for temper, surface treatment, and other technical properties completely unsatisfactory. The explicit goal of the early experiments was to understand if these technical choices affected any utilitarian performance characteristics. Does this mean that this is the only type of knowledge that goes into the manufacture of even the most utilitarian of technologies? Certainly not, nor is this implied in our broader model that was outlined earlier (see also Schiffer and Skibo 1987, 1997; Skibo and Schiffer 2001). Our strategy focused explicitly on utilitarian performance fully realizing that other various nonutilitarian performance characteristics might also play a role in the manufacture and use of this pottery. Our earlier work on pottery was meant to fill the lacunae relating to technical choices and utilitarian performance, not to suggest – as some archaeologists have inferred – that these are the only factors important in pottery manufacture and use.

We chose to look at cooking pots in various contexts because it was a chance to flex our experimental muscle by discerning the effects of seldom-studied technical choices on utilitarian performance characteristics. For cooking pots, we argued, thermal shock resistance is a primary performance characteristic because if a pot has poor thermal shock resistance it will not survive firing during manufacture, much less repeated heating during use. A pot that does not survive long-term boiling episodes will not cook beans, and a family might go hungry. Chapter 3 provides a good example for how this argument works and how we made a case for multifunctionality (cooking, storing, brewing) in vessel design. But many other factors are involved in the adoption of this technology that remain unexplored; perhaps these are the types of questions that practice and agency theorists call for. For example, the adoption of pottery on the Colorado Plateau did not occur simultaneously, and there is a good deal of variability in the technology that is yet unexplained.

Cooking pots turned out to be a rich technology with clear connections between technical choices and utilitarian performance, which could be explored experimentally. What if we had focused, however, on serving bowls, also a common form in the American Southwest America? Although we have not looked at this technology carefully, one can easily envision that only a few utilitarian performance characteristics – such as holding capacity and accessibility of contents – would be of primary concern in their design. Because the pots are used to serve food to family members and guests, it is likely that various sensory performance characteristics (especially visual ones) weigh more heavily in vessel design. A serving bowl is involved in a process of communication between the people taking part in these interactions in manufacture and use activities. The technical choices that went into the design of these vessels cannot be discovered by experimentation alone, because they are greatly affected by various factors relevant to a particular time and place (Sillar and Tite 2000). Someone wanting to explore these choices might start with use-alteration traces on the vessel and then move to the evidence of use across space and through time. These kinds of inferences, however, are more difficult and require much more from the archaeological record. This is why researchers who seek the meaning of artifacts often run into methodological problems and produce unsatisfactory case studies.

According to Pauketat (2001:87), "answers to ultimate 'why' questions will be found only through cumulative, painstaking, data-rich, multi-scalar studies of proximate causation." We could not agree more. The reader will note that there is a distinct difference in the following chapters that deal with modern material culture and those that focus on prehistory. That is because in modern material culture it is far easier to access the types of knowledge that the agency and practice proponents seek. Ancient technology provides a number of difficult hurdles to applying "avant-garde" approaches to archaeology (Pauketat and Alt 2005:232). Our strategy with prehistoric data is to explore the utilitarian performance characteristics first (see also Lemonnier 1992:137). Dobres (2000:36–37) has criticized this strategy, but we should note that it can access some elements of knowledge and at least provide a foundation for moving on to a more complete understanding of a technology's manufacture and use (see also Roux 2003), as is illustrated by Pauketat (2001). In the ball court study (Chap. 6), we do indeed explore several nonutilitarian performance characteristics and argue that they are important for understanding the role of this feature in the thirteenth-century Southwest, yet this line of investigation is not complete. To fully understand the manufacture and use of ball courts requires contextual data of the type that were not available to us. The work of VanPool and VanPool (2003b) and others (Douglas 1995; Newell and Gallaga 2004; Schaafsma and Riley 1999; VanPool 2003; Whalen and Minnis 1996, 2001) is making progress in this regard.

We maintain that agency and practice theory are entirely compatible with the behavioral model, and the two approaches begin to converge when people make serious attempts to apply agency and practice to prehistoric data. It is one thing to engage in what Pauketat and Alt (2005:232) have called "avant-garde" approaches to archaeology, which has characterized much of postprocessual archaeology, but it is yet another to apply these approaches to the realities of the archaeological record. According to Dobres and Robb (2005:161), "archaeologists can only understand agency – and thus social reproduction – when we understand how it worked (and works) materially." To understand how it "worked materially" they advocate the *chaîne opératoire* approach, which for reasons outlined earlier, we believe has serious deficiencies. Dobres and Robb (2005:163), however, suggest that there is a variety of other strategies for "doing" agency including "the life-history approach to material culture and the built environment....(that) focuses on so-called performance characteristics in an attempt to unlock the choices people made in regard to making and modifying their material world." This, of course, is the approach we advocate here.

Dobres and Robb (2005) realize that in order to "do" agency, we must engage the archaeological record in ever more rigorous ways. A number of scholars, such as Cobb and King (2005), Dietler and Herbich (1998), Joyce (2000), Pauketat and Alt (2005), Sassaman (2005), Walker and Lucero (2000), and Van Pool and Van Pool (2003) have made good initial attempts at engaging the archaeological record in ways that will indeed permit us to "do" agency. Pauketat and Alt (2005:230–231) suggest that there are three "procedural fundamentals" for doing agency. First, one must have a firm grasp of archaeological variability through time and space. Second, a researcher should compare "histories of practices," which basically means to investigate a particular technology in its various social contexts of manufacture and use. Finally, they suggest that doing agency requires "tacking back and forth between lines of evidence at multiple scales of analysis." They further suggest that comparisons of this type will derive from "experimental archaeology and studies of natural formation processes, technical performance, and choice." For the latter "procedural fundamental" they refer specifically to our work, but we would argue that this approach is exactly the type of strategy that we advocate and can be seen in the following case studies.

We should note, however, that this is a procedural standard that few have been able to meet. Many, in fact, jump too soon to applying agency even when they lack a firm grasp of the archaeological variability (procedural fundamental number one). Many, including ourselves, have applauded those who have called for an archaeology that investigates all aspects of technology (gender, power, etc.), but often times we have been left unconvinced by attempts to apply these ideas to the archaeological record (see also Killick 2004:575). Few engage the archaeological record in a way that furnishes an adequate understanding of formation processes, which must precede reconstructions of a particular technology (for exceptions, see Chapman and Gaydarska 2007; Shimada and Wagner 2007). This is an exciting period in archaeology as scholars struggle to create convincing case studies tied to the archaeological record and investigate new kinds of relationships between people and things. However, we strongly caution against applying various avant-garde models borrowed from sociology, cultural anthropology, or history directly to archaeological phenomena without significant adaptation. Our position has always been that archaeologists, because of our unique data set and perspective, should build our own method and theory (Schiffer 1975a, Chap. 7). The model we advocate here is created by archaeologists for archaeology because it starts with the basic idea that we are focused on the relationship between people and things. We are, nonetheless, encouraged by the fact that as researchers attempt to apply practice or agency perspectives to archaeology significant overlaps appear between these attempts and what behavioralists advocate. In archaeology – and especially in the study of the distant past – the rubber eventually meets the road; it is at this interface where we see the greatest potential for convergence.

Where the Rubber Meets the Road

If practice and agency are to avoid becoming the next worn-out fad tossed aside by the coming wave of graduate students feverishly scouring the social sciences to find their own niche (Conkey 2007), its advocates must demonstrate its clear connections to the archaeological record. There is nothing wrong with borrowing theory from cultural anthropology, sociology, or history, as long as we frame research questions in terms of people–artifact interactions and establish clear connections between theoretical constructs and the material realities of the archaeological record. In the example that follows, we demonstrate one way that practice theory can be used in archaeology by applying the performance-based life history approach.

We focus on practice theory, as adapted from Bourdieu, because it seems amenable to an easier convergence with archaeology (Dietler and Herbich 1998; Orser 2004:126; Pauketat 2000); yet other agency models, in a general sense, could also be substituted. Before proceeding with the example, we make three points relative to similarities and differences between our approach and practice theory. First, practice theory is more than just *habitus* (Orser 2004:131–138). *Habitus* is "systems of durable, transposable dispositions, structured structures....which generate and organize practices and representations that can be objectively adapted to their outcomes without presupposing a conscious aiming at ends..." (Bourdieu 1990:53). Pauketat (2000:115) notes that *habitus* overlaps with Saussure's *langue* and the concept of "tradition," which is a term familiar to archaeology. Practice creates tradition by "continuous and historically contingent enactments or embodiments of people's ethos, attitudes, agendas, and dispositions" (Pauketat 2000:115). Archaeological application of practice theory, however, "has focused almost exclusively on *habitus*," leading to an incomplete "understanding of social complexity" (Orser 2004:141). In an investigation of race in nineteenth-century Ireland, Orser (2004) suggests that one cannot apply Bourdieu's practice theory without the additional concepts of "capital" and "field."

Capital comes in several forms, including the traditional usage as economic capital, but Bourdieu "extends it to cover all forms of social power" (Orser 2004:133). Thus, there can be cultural capital, social capital, and symbolic capital, which individuals or groups can obtain and use to various ends. Fields are social networks in which the struggle to accumulate capital is played out. For any group, there can be many fields that structure the struggle among actors for the accumulation and the use of social capital.

The second point is that the core concepts of practice theory, *habitus*, practice, capital, and field can be easily accommodated by our model. *Habitus* and practice in our model can be regarded as performance guided by tradition, knowledge, and local contingencies leading to choices by individuals or groups. These choices become routinized to the point where the understanding of a particular practice – for example, the problem it was originally chosen to solve – can be lost and considered just a part of tradition. Every ethnoarchaeologist has had the experience of getting the reply, "Because that's the way we do it," when they inquire about the reason for a particular practice. Our *habitus*, however, is distinctly archaeological and thus material. For us, the routinized activities of everyday life are interactions between people and things.

Field, the network of social relations, is quite similar to our concept of "*cadena*" (Schiffer 2007), which is "all interactors involved in an artifact's life history, both people and artifacts" (Walker and Schiffer 2006:71). For our model to be of use to archaeologists, the *cadena* includes not just the social groups but also the artifacts involved. Another distinguishing characteristic of *cadena*, unlike field, is that it is historical and incorporates the crucial concept of behavioral chain.

The important concept of "capital" is only of use to archaeologists to the extent that it intersects with material culture. In our model, performance characteristics play this role as they are capabilities, skills, or competences that material culture and people must have to perform their functions, whether utilitarian, social, or symbolic. We revisit the Kalinga case study (Skibo 1994) to illustrate the overlap in concepts.

The Kalinga in the late 1980s were using mainly metal (aluminum) pots to cook rice, and ceramic pots to cook vegetables and meat (see also Kobayashi 1994). The women insisted on scrubbing off all of the soot from the exterior of the metal pots so that they would maintain their luster, even though this process was arduous and removed a thin layer of metal. In an earlier study, Skibo (1994) interpreted this behavior as part of the activity associated with a symbolic performance characteristic: the shiny pots were a symbol of wealth and modernization that were proudly displayed in houses. Let us take this simple example and look at it more closely using practice *and* behavioral concepts. The *habitus* of interest here is the routinized pottery-washing activity done by all women and many young girls at least once per day. The washing activity was remarkably similar among all the women observed and left visible traces on the vessels themselves (Skibo 1992). Washing, and the resultant use-alteration traces, varied only by vessel type (rice or vegetable/meat ceramic pots and the metal pots) and the handedness of the washer. The vessels were carried by hand to and from the washing location and after cleaning were placed either on a wooden shelf (ceramic) or hung from the rafters by their handles (metal). The shiny metal pots were hung or otherwise displayed so that they were visible to visitors or guests. The original question was, Why did they go to such bother to shine the pots and then display them in prominent locations? We stand by the answer to that question offered over a decade ago: symbolic performance. This inference can be restated in terms of practice theory and our model, which has evolved since that time.

The field or *cadena* is the set of people from the village of Guina-ang, Kalinga relatives or friends who might visit from other villages, and even more distant non-Kalinga visitors to their house including ethnoarchaeologists. The Kalinga economy at the time was almost completely dependent on subsistence agriculture and there were relatively few overt differences in economic capital. Yet, even within this relative economic equality there were some differences in household wealth measured in terms of number of rice fields, ownership of animals, and house size (Trostel 1994). Guina-ang residents were aware of what we would consider, from the perspective of a western capitalistic perspective, relatively modest differences in household wealth. Added to this is that metal pots were also more expensive than their ceramic counterparts.

Capital takes several forms in this simple case study. There are important differences in economic capital between households, and this is transferred to symbolic capital (or symbolic performance) by conspicuously hanging shiny metal pots in one's house. The primary network of conflict (field or *cadena*) is the other Guina-ang villagers who clearly understand that this overt display of metal pots represents a dominance of symbolic and economic capital (visual performance characteristic facilitating a symbolic function).

This example is an oversimplification but it also clearly illustrates overlaps between practice theory and our model. One important distinction is that our model is materially based and thus more easily applied by archaeologists. More than that, our model makes it possible, at least in principle, to link agency and practice theory rigorously to the archaeological record. We should note, however, that this example pertains to living people, and we are sensitive to the pleas of prehistorians who yearn for tangible ways to explore symbolism and power in the distant past (see Sullivan 2007).

This brings up our third and final point: practice theory or any avant-garde theory cannot be simply mapped onto the archaeological record. Orser (2004:141; see also Orser 2007), in his study of race, notes that Bourdieu's "ideas cannot be used verbatim to understand the practice of race." Likewise, Pauketat (2001:79) notes that, "there is no practice-theory cook book, nor should archaeologists simply reify Bourdieu's concepts as ready made interpretations." We concur completely because Bourdieu's model was based on twentieth-century French society and certainly does not have the material basis required for archaeological application. Bourdieu's theory is also about social power, and so we should confine our archaeological applications, at least initially, to strong cases where power and domination are readily inferred (e.g., Joyce 2000). Pauketat's case study for practice theory is the Mississippian period in the Midwestern USA, where a hike to the top of Monks Mound is all that is required to know that this thirteenth-century society had clear differences in status and social power. Orser's (2004) application of practice theory, done with the aid of textual data, takes place in nineteenth-century Ireland where there were clear differences in social class and power. Although there are differences in social power even in relatively egalitarian societies, as is demonstrated by the Kalinga example, we caution prehistorians about applying practice theory, or any of the trendy postprocessual models, to hunter-gatherer and horticultural societies of the type that dominate prehistoric America until the requisite method and theory are developed (Killick 2004). In that spirit, we offer a final case study that illustrates an artifact-based strategy for exploring the materiality of social power (Walker and Schiffer 2006).

Logging Camps and Social Power

The model is based on the presumption that social power is embodied in the relationship between people and things. The focus is on the difference between structural power, a group's socially defined power, and actual power, which is found in the practices of people as part of a *cadena*. Social power, in this case, is measured by the ability of one group in a *cadena* to acquire artifacts or goods through any number of processes. Choice among alternative artifacts is determined by their anticipated performance characteristics, and because in any *cadena* there can be groups of people with competing agendas and performance preferences, conflict can occur (Walker and Schiffer 2006). For example, the person or group that acquires an artifact or structure may be different from groups that use and maintain that artifact or structure. In using *cadena* to study social power, one focuses first on acquisition events and the person or social unit that has the social power to acquire that artifact. One can then investigate whether other groups were disadvantaged by that artifact.

One of the best places to explore relations between the material and the social is in the historical record, especially in historical archaeology. Establishing connections between people and things among living people can be done in ethnoarchaeology and in historical contexts, but historical archaeology has the advantage of having an archaeological record and textual data. The case study focuses on logging camps that were in use in the Upper Great Lakes during the nineteenth and early twentieth centuries, the pine and hardwood lumber era (Franzen 1992, 1995; Karmanski 1989). There has been little systematic research, including excavation, at logging camps, but they are a particularly rich resource for studying structural and actual social power that is just now being realized (Drake and Drake 2007; Drake et al. 2006; Drake n.d.; Franzen 1995; Hardwick 2008).

Although we use terms from our model (i.e., *cadenas*, performance characteristics, choice), those who employ practice theory should be able to see correspondences to constructs in their framework. Lumber camps have a rather complex social structure made up of many social groups that include: company owners, foreman, auxiliary staff (e.g., cook and aids, blacksmith, animal keepers, mechanics, carpenters), and lumberjacks. We are going to focus on just two of these social groups at the extremes of the social structure: the lumberjacks and the company owners.

We also focus on just two artifacts: saunas and liquor bottles. The *cadenas* consist of the life histories of saunas and liquor bottles and the competing social groups involved. Although the interface of structural and actual social power will eventually be investigated among all the artifacts and social groups at these camps (Drake and Drake 2007; Drake n.d.; Hardwick 2008), the focus here is on two *cadenas* associated with the two artifacts.

In terms of architecture, the company/owner maintained almost complete structural control, purchasing construction materials, organizing the camp layout, and building the structures. The organization of the camp's buildings reflects a number of performance characteristics important to the owner/foreman. The lumber camps were designed as inexpensive temporary structures with very little concern for aesthetic visual performance. Camp foremen were given a sparse budget to build a camp and the emphasis was on utilitarian performance. But because the camps were built for winter habitation they had to be strong enough to withstand the weight of the snow and offer enough protection so that above-freezing temperatures could be maintained in the living quarters.

Although logging was a dangerous occupation and many men were killed or maimed on the job, it was in the company's best interest to have the loggers productive and free of illnesses that would keep them from doing their work. Thus, kitchen design and procedures were meant to be as sanitary as possible, outhouses were built, and fresh food and water were provided. The company was concerned with health only as it affected productivity, but as we see later, ethnicity-based conceptions of cleanliness led to a disjunction between actual and structural power as seen through artifact acquisition.

Many camps were built with military precision that clearly demarcated, architecturally, the social structure of the camp. The loggers were housed in a single bunkhouse and the foreman (on-site representative of the owner) often lived in a separate cabin (Drake and Drake 2007).

The structural social power differences between owners and loggers are clearly evident in acquisition events related to camp architecture. The *cadena*, then, consists of the groups involved in the life history of camp structures. The owner did all the purchasing of materials for the camp and hired laborers to do the construction. Performance characteristics weighted heavily were ease and cost of construction while maintaining a minimal level of comfort for the workers. Although the camp was deficient in visual aesthetics – many of the camps were the quintessential tar-paper shacks – that does not mean that visual performance characteristics were unimportant. The layout of the camp and the structures themselves symbolized the greater power of the owner and the subordinate status of the workers (Drake and Drake 2007).

One advantage of archaeology is that it provides an opportunity to investigate differences between structural and actual powers at the camp, and this is most evident in a special type of building, the sauna, found at many camps in the Upper Peninsula of Michigan (Drake and Drake 2007; Drake n.d.; Franzen 1992:90–91). The sauna was brought to the region by Finnish immigrants beginning about 1890, followed shortly by a number of immigrants of eastern European descent, such as

Poles, Slovenians, and Croatians (Franzen 1992:83–84). The Finns were unique, however, in that they insisted in many cases that logging camps be furnished with a sauna, which for them was important for health and sanitary reasons, or what we would refer to as utilitarian performance characteristics. It is clear, however, that the sauna, which was more of a communal ritual done weekly rather than a sanitary necessity, had symbolic performance characteristics that served to maintain ethnic cohesion (Drake and Drake 2007). So here we see that despite lacking structural power, the Finnish loggers exercised actual power over the acquisition of a type of architecture, the sauna.

As Drake and Drake (2007) note, there is evidence for both the logging company building saunas, though they deducted building costs from logger's wages, and individual loggers constructing the structures themselves on their own time. The important element is that loggers, who do not participate in any acquisition events for camp buildings, did control in some cases the acquisition and construction of the sauna. Given that the saunas were the traditional *savu* or smoke sauna, which are very simply constructed (Drake and Drake 2007; Franzen 1992:90), it was possible for Finnish loggers to build the structures themselves when they were not on "company time." The acquisition events, therefore, are the after-hours labor where the loggers acquired the materials from the local environment (rocks and timber) or scavenged it from the company. The construction of the sauna is the only camp structure acquired by the loggers, which gave them some social power in an environment of domination. It is no surprise, therefore, that Finnish loggers were behind the 1936–1937 timber worker's strike, which emphasized living and working conditions instead of wages (Franzen 1995:330). Finnish socialist and labor associations were behind the strike, which led to the end of the company logging camp (Franzen 1992:26). Eventually, the loggers gained complete control of acquisition of materials at logging camps as family-based operations became much more popular than the company camp.

A final logging camp example illustrates how social power is defined by acquisition of alcohol (Franzen 1995) and the difference between structural power and actual power. Structurally, the company provided all material possessions needed by loggers for the extended stays in the isolated camps. Food was furnished and any personal items were supplied by the company at the "van" or commissary. Here the logger could purchase items such as tobacco (for smoking and chewing) and other items, but the van did not offer alcohol, and the company banned its consumption. Various patent medicines, such as Hinkley's Bone Liniment and Dr. Kilmer's Swamp Root, were sold at the van and the "medicines'" primary ingredient was often alcohol (Franzen 1995:301). Most of the loggers wanted to drink alcohol or ingest the patent medicine but the company tried to keep consumption in check by selling the medicine in the vans where the keepers could control distribution (Franzen 1995:309). From the perspective of the company, alcohol consumed as either medicine or in the more traditional form could disrupt camp life and hurt worker productivity, and so alcohol was universally banned by the camps and medicine consumption was controlled. The workers had only marginal control of alcohol acquisition, which demonstrates the company's tremendous social power.

Archaeological evidence from logging camps, however, suggests that alcohol was consumed despite being banned (Franzen 1995). Preliminary testing at the Underhill Camp on Grand Island, a company camp operated in the early twentieth century (Hardwick 2008), found whiskey bottles around structures and in the privy. Franzen (1995), looking at surface finds at a number of logging camps, found evidence of alcohol bottles, which were likely discretely tossed out into the snow while the camp was operating.

The loggers increased their actual power by acquiring alcohol elsewhere, sneaking it into camp and then drinking secretly. What performance characteristics did the alcohol possess? Like the sauna construction, we can identify both utilitarian and symbolic performance. Clearly, the alcohol (either as whiskey or patent medicine) would provide a utilitarian function. Logging was hard, cold, and dangerous work resulting in many serious injuries. All loggers after just a few days in the woods would have to be working with a variety of minor injuries and discomforts, as well as the emotional difficulties that might have come about by living in an isolated camp away from families (Franzen 1995:328). Alcohol, either secreted into the camp or purchased at the camp van in the form of patent medicine, would have been a way of self-medication and certainly assuage these physical and emotional ailments.

The symbolic performance characteristics of alcohol purchase and consumption have to do with the ethnic attitudes toward alcohol brought to the camps by the diverse groups. For many eastern European groups, also represented at the camp, alcohol consumption was a part of daily life that they wanted to continue in the camps. Ironically, many of the Scandinavian groups, including the Finnish sauna builders, were nondrinkers.

These examples should be considered a preliminary exploration of social power at logging camps, which have a relatively simple organization, yet are composed of a complex set of power relations. Nonetheless, this case study illustrates how the acquisition model can be applied to investigate structural and actual power, and how the historical archaeology of logging camps can serve as a good testing ground for understanding social power. Structurally, the company had near-complete control of both the camp buildings and alcohol consumption, as demonstrated by who controlled the acquisition. Actual control, however, can be seen in the archaeological record as Finnish loggers built saunas, and other loggers were able to drink alcohol even though it was banned in the camps.

Conclusion

Returning to the Kalinga household introduced at the beginning of the book, the Kalinga man had minimal knowledge of pottery because he was not, like the women, immersed on a daily basis in the use of this technology. The female pottery users made choices regarding which pots to use for rice and which for vegetables, the size of the pot, and how to cook various items based on their knowledge, experience, and traditions. They chose which vessels to acquire based on performance characteristics related to the quality of the pots and their social relationships with the potter (Aronson et al. 1994). Some vessels in the rafters were heirlooms passed down from their grandmother and mother, and each of these pots had important meanings to the user. In terms of social power, women controlled completely the acquisition of pots, which illustrates that women did have some social power in the household. But because the artifact that they controlled, pottery, was relatively insignificant economically among the Kalinga, their power was tempered as men controlled the acquisition of more valued commodities.

Just as the Kalinga man could learn these things about his wife's technology, so can archaeologists begin to unravel this sometimes complex relationship between people and artifacts. It is not easy to do even in the simplest technology, but it can be done. In the model presented and in the case studies that follow, we offer our perspective for understanding the relationship between people and things.

Chapter 3
The Origins of Pottery on the Colorado Plateau[1]

In the days of Gordon Childe (1951), the emergence of pottery seemed sudden and easily understood. Sedentary agriculturalists made pottery, and it signaled the beginning of the Neolithic revolution worldwide. Although this is still generally true, more recent research and better dating techniques have made this once simple equation between pottery and sedentary agriculturalists much more complicated (Pavlů 1997; Rice 1999). We now know that mobile hunter-gatherers made pottery (e.g., Aikens 1995; Bollong et al. 1993; Reid 1984; Sassaman 1993; Tuohy and Dansie 1990), and some cultivators, like those of the Lapita Culture (Green 1979), actually abandoned pottery technology. In areas such as the American Southeast, pottery manufacture preceded agriculture for up to 2,000 years, and in the American Southwest or the Near East agriculture was present long before the first pottery.

In this chapter, we first examine the origin of pottery generally, and then look more closely at one particular case – the emergence of pottery on the Colorado Plateau of the Southwestern USA. The analytical focus of this study is a sample of whole and partially reconstructed vessels from sites dating between AD 200 and AD 600. Using a performance-based analysis, the functions of the early vessels are inferred through an analysis of morphological characteristics and use-alteration traces. The collections of whole brown ware vessels from three sites in northeastern Arizona are dominated by globular neckless jars. From a performance perspective, it is argued that these vessels would have performed very well as storing, cooking, or processing vessels. Preliminary use-alteration analysis suggests that some of the vessels were not used over a fire, whereas others were used in two types of cooking. Moreover, many of the vessels were used for alcohol fermentation that caused extreme interior surface attrition.

Origins

The oldest ceramic objects in the world thus far are the Dolní Véstonice figurines that date to about 26,000 years ago (Vandiver et al. 1989), preceding the appearance of pottery *vessels* by over 15,000 years (see Pavlů 1997; Rice 1987:6–16, 1996a,

[1] This chapter is cowritten by Eric Blinman

1999 for a general reviews of pottery origins). What concerns us here is not the initial invention of ceramic technology, but rather the innovation of ceramic containers. Most archaeologists would now agree that long before the widespread adoption of pottery, hunter-gatherers had knowledge of the basic principles of ceramics: objects can be shaped from moist clay and then be made permanent by placing the object in a fire (Brown 1989:207; Rice 1987:7). The issue is when, where, and why pottery containers make their appearance, and it is clear that there is no single answer (see Arnold 1999b; Barnett and Hoopes 1995; Vitelli 1999).

Although there may not be one reason for the adoption of pottery containers, Arnold (1985) identifies a number of generalizations about pottery and people based on both ethnographic and archaeological observations. The two of most interest here are the relationship between pottery making and sedentism, and the correlation between pottery and more intensive forms of food processing.

Nonsedentary and semisedentary peoples can and do make pottery, but Arnold (1985:113–118) found a strong correlation between pottery making and sedentism. There are several reasons why this would occur. Pottery is less portable and more prone to breakage than other containers such as baskets and skins. Although this may seem to be a logical reason for the lack of pottery among mobile peoples, in practice it may have been only a minor impediment (see also Arnold 1999b). Some hunter-gatherers do carry their pottery vessels with them (e.g., Holmberg 1969; McGee 1971; Sapir 1923), and sedentary people often transport their pottery over long distances (Arnold 1985:111). A more important reason behind the correlation between pottery and sedentism is that pottery making is a technology that takes some investment (Arnold 1985). Although clay is somewhat like McDonald's hamburgers, in that you can always find some nearby, the nearest available clay may not be appropriate for particular pottery-making techniques. For example, locally available alluvial clays may be inappropriate for vessel manufacture because of excessive shrinkage. Among contemporary potters you find that once a good clay source is found it may be exploited for generations because of its known and acceptable working properties (Reina and Hill 1978). People with a mobile settlement and subsistence system may find it difficult to establish and maintain a pottery technology if they do not at least have access to the same pottery resources on a yearly basis. As Brown (1989:116) notes, at least seasonal sedentism may be required for pottery manufacture.

The final reason why sedentism is important to pottery making is because of scheduling conflicts (Arnold 1985:99–108; Crown and Wills 1995). Potters must be near a good clay source during a season of the year when potting is possible and when they have time, free from other tasks, to make pots. In many parts of the world, pottery can only be made during one season of the year because of climatic restrictions (e.g., too wet or too cold), and so scheduling conflicts can indeed be an impediment.

The second generalization made by Arnold (1985:128–144) relates to pottery vessels as tools for food processing. Pottery sherds are the most ubiquitous artifact found at Neolithic or Formative villages worldwide because ceramic vessels had become an essential tool for the processing of staple cultigens, allowing high temperature processing for long periods of time. Boiling or near-boiling temperatures are essential for making many foods palatable and digestible. Cereal grain starches must be completely gelatinized for maximum digestibility, which requires sustained temperatures over 93°C (Reid 1990:10; Stahl 1989:181). Boiling, steaming, or simmering can also destroy potentially harmful bacteria and eliminate or reduce toxins in cultigens (Arnold 1985:129–134; see also Stahl 1989:182). Moreover, cooking in pots can increase the nutritive value of meat (by extracting fat from bones) and some leafy vegetables (Reid 1990).

Compared with other cooking containers, pottery vessels permit direct heating with less constant attention. Although indirect heating of water with hot rocks (as in basket boiling) is an effective way to reach boiling or near-boiling temperatures, it requires continuous attention to avoid boilover and to maintain those temperatures for long periods of time. When ceramic containers are used, once the relationship between the heat source and the pot is established (nestled in coals, supported over the fire, etc.), constant temperatures can be maintained by occasionally tending to the fuel. Ceramic vessels also provide sturdy processing containers for preparation techniques such as fermentation or alkaline soaking that may break down other types of containers. Clearly, ceramic containers provide many advantages as cooking and processing tools, permitting the exploitation of many new foods and the more effective processing of others (see also Crown and Wills 1995:245–246).

Cross-cultural generalizations can provide insights into the relationship between pottery and people and shed light on ceramic vessel origins, but these data cannot be applied simply to explain pottery origins. To search for the clues to specific pottery origins we must turn to the archaeological record.

Rice (1999) and Barnett and Hoopes (1995) provide a good worldwide summary of some of the earliest pottery technologies, and it is clear that there is not just one explanation for pottery origins. The striking aspect of early ceramics is the lack of correlation between pottery making and agriculture. Although pottery becomes the processing workhorse for agriculturalists, as described earlier, the earliest people to use pottery as a tool were hunter-gatherers. In many parts of the world, it was hunter-gatherers who first employed ceramic containers to process food. Indeed, the earliest known pottery vessels in the world are small cooking pots that come from Fukui Cave on Japan's southernmost island (Aikens 1995). Incipient Jomon pottery, as it is called, appears on sites with evidence of intensive marine harvesting during the Pleistocene–Holocene transition beginning about 12,400 b.p. (uncalibrated).

In North America, there are many examples of hunter-gatherer pottery, mostly in the southeastern and northwestern USA, but extending into Canada and Alaska as well. There is evidence that these pots were also used as processing tools (Reid 1990; Sassaman 1993, 1995). The majority of these vessels are low-fired open-bowl or jar forms often tempered with organic matter. Although these Late Archaic vessels often have soot on the exterior suggesting that they were used over a fire (Beck et al. 2002; Sassaman 1993), both Reid (1990) and Sassaman (1993, 1995) make the argument that these vessels may have been used to process food by indirect moist cooking (i.e., stone boiling) as well. The highly porous thick walls and open mouth make poor heat conductors but excellent insulators, which is a performance characteristic that would be well suited to simmering foods by indirect heating. They argue that simmering temperatures, easily maintained by indirect heating, were employed by these hunter-gatherers to stew meat and obtain oils from seeds and nuts or the marrowfat from bones (Reid 1990:10; Sassaman 1995).

But processing of food cannot explain every case of early pottery. In some regions of both the Old and New Worlds, the earliest ceramic vessels were not tools for food processing but rather were important artifacts of ritual activity. The early pottery of Colombia is highly decorated, and Oyuela-Caycedo (1995) argues that these vessels were not used for cooking. Clark and Gosser (1995:116) also suggest that early Mesoamerican pottery may not have been used for food preparation. In the Old World, Vitelli (1989, 1995, 1999) also finds that early vessels of the Greek Neolithic were not used for cooking, and she suggests that these early assemblages played a symbolic or shamanistic role.

To summarize, early pottery around the world appears in three separate contexts: (1) sedentary cultivators that use the vessels to process and make digestible cereal grains, (2) seasonally sedentary hunter-gatherers who use vessels with either direct or indirect heating to extract additional nutrients from animal products or to more effectively process seeds and nuts, and (3) early cultivators or hunter-gatherers who produce and use the vessels in ritual activity. The first two contexts involve food processing and are much more widely documented than the evidence for the ritual use of pottery. The latter context will be better understood after more information is gathered on vessel use.

Theoretical Models

Several scholars have attempted to explore the origins of pottery from a theoretical perspective. We will review the models proposed by Brown (1989) and Hayden (1995) as they may be the most relevant to the origins of pottery on the Colorado Plateau (see Rice 1999 for a thorough review of these and other models).

Brown (1989) revived interest in the origins of pottery by exploring an economic approach. His model considers that (1) pottery containers were adopted long after there was knowledge of ceramic technology, (2) pottery was introduced when people had other well-developed container options, and (3) pottery is not the only container for heating water and processing food (Brown 1989:208). Under these conditions, pottery was used when there was a "rising demand for watertight, fire-resistant containers…coupled with constraints in meeting this demand" (Brown 1989:113). In this model, groups would have to be at least seasonally sedentary to permit pottery to be a realistic container option. Pottery is adopted when other types of containers such as baskets or skins fail to meet the increasing demand brought about by new types of food processing, new forms of storage, or the emergence of food presentation as a form of social expression (Brown 1989:113). Thus pottery was not used because of some foreseen potential but rather because it was a container that could be made cheaply and quickly by semisedentary groups.

Hayden (1993, 1995) looks at prehistory and does not see people trying to solve the practical problems of life, but rather he sees individuals involved in economically based competition. As in Brown's model, prerequisites for the emergence of pottery are technological advances and more sedentary settlement and subsistence systems. Hayden (1993) argues that as people become more sedentary and sharing of food is no longer required for survival, there is a worldwide tendency for increased economic competition along with more pronounced inequality. In this context, pottery first appears as a prestige food container made by individuals in direct competition with their neighbors.

The primary difference between the Brown and Hayden models is the role of practical versus prestige technologies. Although they both are economic models, Brown suggests that the demand for pottery containers was to fulfill practical needs, whereas Hayden promotes the idea that demand for pottery was generated by economic competition. The implications are that Brown's model predicts that the earliest pottery in a region would have been processing vessels, whereas Hayden's model predicts that the first pottery would have been food-serving containers. As noted earlier, both situations can and do occur worldwide. Some researchers have found that the earliest pottery in a particular region was used to cook or process food (e.g., Gebauer 1995) and others have shown that the first ceramic containers, often highly ornate, were not used in food processing but, presumably, as a prestige technology (e.g., Clark and Gosser 1995:2.14–2.16; Oyuela-Caycedo 1995).

These models are not mutually exclusive. Although Brown (1989) focuses principally on practical demands as an impetus for pottery and Hayden (1995) suggests that social or economic competition was the important factor, they each leave room in their models for the opposite to occur. Brown (1989:113) notes that one of the new container demands could be the "presentation of food as an emergent social expression." Similarly, Hayden (1995:261) suggests that in the process of producing pottery as a prestige good, its practical benefits are quickly realized and put into use. Moreover, in some peripheral areas, "derivative practical pottery" used for cooking or storage may have been the first ceramic vessels (Hayden 1995). Clearly, there is a great deal of overlap between the two models, with the main difference being the weight placed on prestige versus practical ceramic containers. It is possible that each can be used to explain the emergence of pottery in various parts of the world, but testing the models requires a level of analysis that is rarely attained. What is often lacking is a clear idea of how the earliest pottery was used (Longacre 1995; Rice 1999). The example that follows attempts to remedy this deficiency with an analysis of the earliest pottery on the Colorado Plateau.

Emergence of Ancestral Pueblo Pottery

Ancestral Pueblo pottery is known worldwide for the elaborate forms, made without the help of the wheel, and its intricately painted designs. If you consider prehistoric North American pottery traditions from the perspective of art, Ancestral Pueblo pottery is at the top. And from the perspective of the Southwestern archaeologist, no single artifact class has played a more important role. From defining culture groups and marking the passage of time, to inferring population size and social organization, pottery from the Colorado Plateau is usually at center stage. But despite the attention paid to this artifact type and the important role it plays in archaeological inference, very little attention has been given to the origins of this pottery (for exceptions see Crown and Wills 1995; LeBlanc 1982).

This scant attention is not for lack of collections since much of the early ceramic material we will describe was excavated decades ago. But we can identify several reasons for this lack of interest. First, it is only recently that we have better data on important issues related to pottery origins, such as the appearance of cultigens and beginning of more sedentary settlement (Crown and Wills 1995:241). Without understanding these important covariables, pottery emergence is not easily explained. Second, the earliest pottery on the Colorado Plateau is brown, and every introductory student in Southwestern archaeology knows that Ancestral Pueblo pottery is gray, and Mogollon pottery, located just southeast in the mountain transition, is brown. Prior to more accurate dating of the brown ware sites, it was often assumed that the brown pottery was imported from the Mogollon region or represented Mogollon immigrants. Third, dates for the early brown ware pottery are consistently prior to AD 600, thus placing it in the Basketmaker II period. Generations of Southwestern archaeologists were taught that there was no pottery during the Basketmaker II period. Although in the Southeastern U.S. archaeologists have come to accept that there is Archaic pottery, the time-honored Pecos Classification has indeed served as an impediment to studying the earliest Southwestern ceramics.

In the Southwest, as well as in most parts of the world, there is evidence that people were well aware of ceramic technology long before the manufacture of pottery containers (Crown and Wills 1995:244). Unfired clay figurines that date between 5600 and 5000 BC have been found in southeastern Utah (Coulam and Schroedl 1996), and ceramic figurines have been located in a southern Arizona pithouse village that dates to about 800 BC (Huckell 1990). It is safe to assume that Archaic people throughout the Southwest had knowledge of ceramic technology. Domesticated cultigens also preceded the appearance of pottery vessels, which is analogous to the Near East and the prepottery Neolithic. Corn was introduced into a mobile hunter-gatherer subsistence system by at least 1000 BC (Tagg 1996), followed by an apparent transition to a more logistic settlement system with semisedentary occupation of pit structures in rock shelters and camps (Crown and Wills 1995; Matson 1991; Wills 1988). More than a millennium later, pottery appears to have been used on a regional scale over the course of one or two centuries, accompanied or closely followed by the architectural and material correlates of the Hohokam, Mogollon, and Ancestral Pueblo (Crown and Wills 1995; LeBlanc 1982).

On the Colorado Plateau of Arizona, New Mexico, Utah, and Colorado, there is now widespread, though scattered, evidence that the first pottery was made sometime before AD 300 (see Wilson and Blinman 1993, 1994, 1995; Wilson et al. 1996).

The pottery occurs in contexts that are similar in all respects to aceramic settlements of the same time. This pottery, known regionally as Los Pinos Brown, Sambrito Utility, Lupton Brown, Adamana Brown, Obelisk Utility, and Obelisk Gray, is a plain polished brown ware (Spurr and Hays-Gilpin 1996; Wilson 1989). In most of the cases, the pottery appears to be locally made (although this must be confirmed with subsequent testing), and in all cases it precedes the typical gray and white wares. A similar stage of incipient pottery manufacture was identified by Haury (1985) to the south in the Mogollon area and in the deserts of the Hohokam homeland (Heidke et al. 1997). Although there is a good deal of regional variability, this early brown ware represents a pan-Ancestral Pueblo ceramic tradition made with self-tempered alluvial or soil clays that tend to be rich in iron. All of the vessels were made using the coil and scrape technique with the possible exception of Adamana Brown, some of which may have been finished using a paddle and anvil (Mera 1934). All of the early brown wares have polished exteriors and surface color ranging from dark gray to brown (for detailed descriptions see Spurr and Hays-Gilpin 1996; Wilson and Blinman 1993; Wilson et al. 1996).

Early Ceramic Sites

Early brown ware sites are currently known from three areas of the Colorado Plateau: (1) the eastern portion of the northern San Juan, which includes the Upper San Juan, Animas, La Plata, and Mancos river drainages, (2) the Prayer Rock District on the Navajo Reservation in northeastern Arizona, and (3) along the southern portion of the Colorado Plateau from the Petrified Forest to the Zuni Reservation. Other sites with this early pottery include the Little Jug site (Thompson and Thompson 1974) near the Grand Canyon, the Hay Hollow site (Martin and Rinaldo 1960), a site east of Gallup, New Mexico (Blinman and Wilson 1994), and a number of sites in Chaco Canyon (for a review of early pottery sites see Breternitz 1982; Fowler 1991; LeBlanc 1982; Morris 1927; Schroeder 1982; Wilson et al. 1996).

An early ceramic period occupation was identified in the northern San Juan area of northwestern New Mexico as part of the Navajo Reservoir archaeology project (Dittert et al. 1961; Eddy 1966). Eddy referred to the earliest pottery as Los Pinos Brown. Although the Los Pinos sites with pottery are not well dated (Eddy 1966:444–445), the pottery clearly pre-dates the later gray wares and represents the earliest attempt at pottery manufacture in this region. Sambrito Brown, which follows Los Pinos Brown in time and is indistinguishable from this type (Wilson and Blinman 1993), provides a larger ceramic sample and comes from slightly better dated contexts (i.e., AD 400–700).

Sites in the Petrified National Forest may represent the best collection of pre-AD 300 brown ware pottery on the plateau. Excavations at the Flattop site (Wendorf 1953) and Sivu'ovi (Burton 1991) yielded a plain brown pottery type classified as Adamana Brown (Mera 1934). Recent dates from the two sites (Burton 1991:97–101) suggest that Adamana Brown may be the oldest dated pottery on the plateau.

The caves of the Prayer Rock District of the Navajo Indian Reservation provide evidence of early pottery making in the Southwest (Hays 1992; Morris 1980). The caves yielded both a classic Basketmaker III pottery assemblage and an earlier assemblage dominated by a pottery type that is called Obelisk Gray. Obelisk Gray is a polished brown ware that is similar to the brown wares described earlier (Wilson and Blinman 1994).

This chapter demonstrates that pottery manufacture was taking place on the Colorado Plateau after AD 200. There is also strong circumstantial evidence that the pottery is locally made, not "Mogollon," and thus not imported from south of the Colorado Plateau (Burton 1991:108; Eddy 1966:384; Fowler 1991; Wendorf 1953; Wilson and Blinman 1993:16). Because similar pottery types are not made in the Mogollon region, we must be careful to distinguish ceramics of the Mogollon tradition from brown ware technology, per se (see Fowler 1991). Many alluvial clays and some geologic clays will fire to brown colors, so the similarities between Mogollon brown wares and those of the Colorado Plateau may represent a similar technology in the first attempts at pottery manufacture (see also Wilson 1989; Wilson and Blinman 1993, 1994).

The Study

The project involved both an analysis of whole vessels and a preliminary clay resource survey from the Petrified Forest area of Arizona to the vicinity of Crownpoint, New Mexico. The objective of the study was to both understand why people started making pots at this place and time, and why the technology changed so rapidly to the typical gray wares.

Initial laboratory analysis focused on collections of whole vessels curated at the Arizona State Museum and Western Archeological and Conservation Center in Tucson, and the Museum of Northern Arizona in Flagstaff. Several vessels from the Laboratory of Anthropology in Santa Fe, New Mexico, were also inspected. These vessels were analyzed and the formal characteristics were recorded to draw inferences about their *intended* function. We also recorded the use-alteration patterns of interior carbon and exterior soot deposits, as well as attrition in an effort to determine *actual* vessel function.

The whole and partially reconstructed vessels come from three sites: Flattop, Sivu'ovi, and the Prayer Rock Caves. Sivu'ovi is located in the Petrified National Park, about 20 miles east of Holbrook, Arizona. The site is a large Basketmaker period pithouse village that was partially excavated by the National Park Service archaeologists to salvage material that was eroding off the small mesa (Burton 1991). The pottery consists of 4 restorable vessels and 1,072 sherds that were recovered from the surface and from 2 pit structures. The vast majority of the ceramics are an early brown ware referred to as Adamana Brown. Similar to all the other early brown wares, it is lightly polished and tempered with fine sand that may be naturally occurring within the clay source or may be augmented by the potter (Rye 1976).

The distinguishing feature of Adamana is the presence of mica inclusions in the temper (Shepard 1953).

Within sight of Sivu'ovi is Flattop, another site dominated by Adamana Brown pottery. Wendorf (1953) excavated 8 pit structures at Flattop and recovered 30 whole or restorable vessels and 2,522 sherds, with all but 84 classified as Adamana Brown. Wendorf did not obtain absolute dates, but ceramic cross-dating suggested that the site pre-dated to AD 500 and was contemporaneous with the earliest Mogollon ceramics (Wendorf 1950:49, 1953:51–53). For example, Adamana Brown was the most common intrusive in the Hilltop phase (tree ring dated to AD 200–400) at the Bluff site (Haury 1985). Burton (1991) obtained radiocarbon dates from two Flattop houses and three houses from Sivu'ovi that confirmed Wendorf's suspicion that Adamana Brown pottery dates very early. Multiple samples were obtained from outer rings of construction timbers, and calibrated dates were averaged for each structure. Burton (1991:101) reports the dates as follows (one-sigma range): Flattop House D, AD 130–318; Flattop House H, AD 35–215; Sivu'ovi Structure 1, 86 BC to AD 131; Sivu'ovi Structure 2, AD 82–252; and Sivu'ovi Structure 3, 406–311 BC.

The caves in the Prayer Rock District of the Navajo Nation were excavated by Earl Morris in the 1930s, and Elizabeth Ann Morris (1980) prepared the report of the excavations and artifacts. Our analysis focuses on the Prayer Rock Caves material because it is one of the largest collections of early Basket-maker pottery. Although the majority of whole vessels come from the slightly later gray ware period, there are also a significant number of brown ware whole vessels and sherds referred to as Obelisk Gray (Morris 1980). This is a bit of a misnomer because this type is quite comparable with early brown wares found elsewhere in the Southwest (Wilson and Blinman 1993; Wilson et al. 1996).

Whole Vessel Design and Performance

There are a total of 211 whole or partially reconstructible vessels from the Prayer Rock Caves, and 74 of those are Obelisk Gray. The remarkable aspect of the Obelisk Gray collection is that half of the vessels are globular neckless jars (Table 3.1), which in Southwestern vernacular are referred to as "seed jars" (this shape is almost identical to the Mesoamerican *tecomates*). Three out of the four whole vessels from Sivu'ovi were also seed jars, and the most common restorable vessels from Flattop were the globular jars without a neck. The early brown ware seed jars

Table 3.1 Obelisk gray vessel forms from the Prayer Rock Caves curated at the Arizona State Museum

Seed jars	37	50%
Necked jars	33	44.6%
Pitchers	2	2.7%
Total	74	100%

are generally spherical in shape, although some are more elongated. They are relatively thin walled and have a restricted orifice. The exteriors, however, are what make these seed jars and all the early brown wares unique. The exterior surfaces are typically quite irregular but they all show evidence of polishing. Sometimes the polish is only visible on the high points of the surface, whereas in other cases more time and effort has been put into smoothing and polishing, resulting in relatively lustrous surfaces.

Based on these technical properties alone, one can begin to make general inferences about intended vessel function and performance. The globular shape of these vessels is a very strong structural design that would impart strength in both the manufacturing and use stages. Shapes approaching spherical have the most green-strength and would be more likely to survive drying without cracking. This would be especially important if alluvial clays of differing shrinkage characteristics were being used within the brown ware tradition, allowing the potter to achieve successful results with either low- or high-shrinkage clay. The same spherical properties also would give the vessel a good deal of strength in use. Curved surfaces have greater structural integrity and thus can better withstand the strains imposed by both thermal shock and physical impact. Moreover, spherical shapes are better able to distribute the weight of their contents, reducing the risk of breakage from internal loading.

The restricted orifice diameter imparts a number of techno-functional qualities. In the seed jar shapes, the strength of the pot increases as the orifice diameter decreases. The small openings are easily covered or plugged to protect the vessel's contents. Moreover, even if the vessel were left uncovered, the restricted opening would limit loss of heat during cooking or spillage during transport or storage. But the restricted orifice also limits access to the vessel's contents. Although all of the analyzed seed jars had openings large enough to permit the entry of a hand or ladle, these openings were small enough to inhibit both access and visibility. Even with lamps for analysis it is difficult to inspect the interior of the vessels, and with a hand or implement in the opening it would have been impossible for the vessel users to see the pot's contents. Moreover, this type of opening is not well suited to pouring liquids, which would not only be difficult to control but would also slop onto the sides of the vessel.

Polishing or burnishing is usually associated with decorated wares in the Ancestral Pueblo Southwest, but it is a technical property that can also greatly influence performance. One of the most important performance characteristics of polishing is its effect on water permeability (Schiffer 1988a). In low-fired earthenwares, water permeability is a constant concern. Without any surface treatment to impede permeability, most vessels will weep badly and greatly reduce heating effectiveness. In fact, water will not boil in some low-fired pottery without a surface treatment to at least slowdown water permeability (Skibo 1992:165–168). But polishing is not often a property found in low-fired cooking pots because escaping water turns to steam and will spall the surface (Schiffer 1990; Schiffer et al. 1994b). This may be the reason for the "poor" polishing job on the early brown ware vessels. They are polished just enough to inhibit the flow of water, but the surface is open enough to permit the escape of steam.

The technical properties of these seed jars, when combined, create vessels that would perform well in both cooking and storage (see also Arnold 1999b). The two most important performance characteristics of cooking with water are thermal shock resistance and heating effectiveness. The spherical shape, thin wall, low firing temperature, and large amounts of temper create a vessel with excellent thermal shock resistance. The thin walls and high percentage of temper also provide excellent heating effectiveness. The polished exterior would also inhibit the flow of water, which is an important property related to heating effectiveness, possibly without closing the exterior surface enough to cause steam spalling. Thus, from a design perspective, the seed jar forms would perform well as cooking pots. The only property of these vessels that is not well suited to cooking is the restricted access. The narrow openings would give the vessels greater strength but also make it more difficult to access the vessel's contents.

As a storage or processing vessel, the seed jar forms also would perform adequately. The spherical shape is a design well suited to storage because of its strength both in terms of holding heavy contents and in being carried while full. Moreover, its low center of gravity, despite its spherical shape, makes it quite stable while resting on its base. The restricted vessel entry is also easily plugged to protect the pot's contents, but it would not be the best design for a storage pot that needs to be accessed regularly or one that requires that its liquid contents be poured out.

From a purely design perspective, the early brown ware seed jars could have adequately performed cooking, storage, transport, or food processing. These designs are multifunctional, and if a person wanted a pot to perform many different functions, the early brown ware seed jars would be ideal. The globular neckless jars with the paste characteristics and surface finish of the early brown wares are the ceramic equivalents of Swiss army knives – one tool that can perform a variety of functions (see also Schiffer Chap. 7).

Whole Vessel Use-Alteration Traces

The majority of analyzable seed jars are Obelisk Gray examples from the Prayer Rock Caves collection. Unfortunately, most of the vessels inspected came from burned houses, which greatly hindered our ability to infer use from carbon deposition. A total of 26 of the 37 seed jars inspected had evidence that postuse burning significantly affected both interior and exterior carbon patterns. Only seven of the vessels survived the burning without evidence that their carbon patterns had been altered. House fires of the type at Prayer Rock Caves can either add or remove carbonized deposits. Fortunately, carbon patterns from the house burning could be easily discriminated from those created during cooking over an open fire. Of the seven pots unaffected by the house fires, two had evidence of cooking and five had no evidence that they were placed over a fire. Both cooking pots had exterior sooting patterns characteristic of being placed over the fire on rocks or on some form of support. The interior of one of the vessels (ASM 14313) had a carbon pattern typical of vessels that heat food in the absence of water (Fig. 3.1). This can occur by roasting seeds or some other food, or by boiling something until all or most of the water has been removed. Cooking a thick gruel would also create this pattern, as would reheating previously cooked food. The other vessel (ASM 14400) has an interior carbon pattern more typical of cooking food in the presence of water (Fig. 3.2). The base has no evidence of carbon while the middle interior has a ring of carbon. When you boil with water, organic particles spatter from the water surface, adhere to the vessel wall, and carbonize. This vessel has a wide ring, as if this pot was used with various water levels or in cases where the water level had boiled down during use.

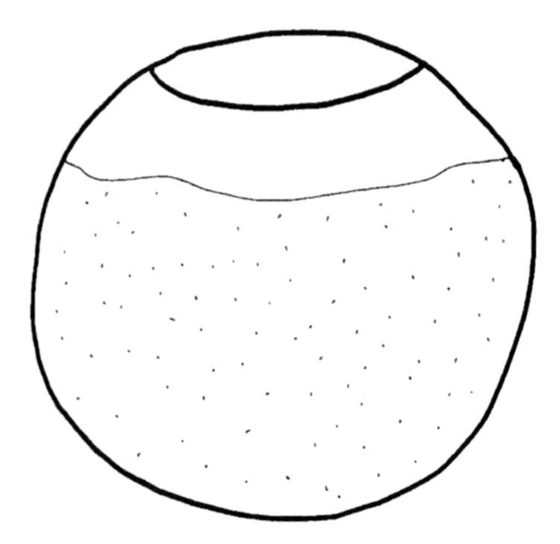

Fig. 3.1 Interior of vessel with a carbon pattern caused by heating food in the absence of water

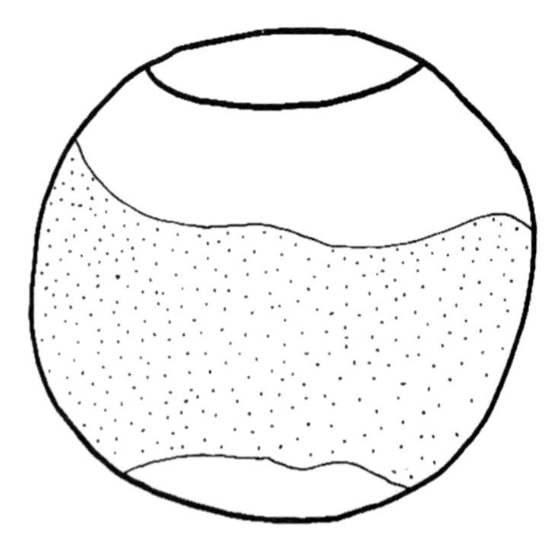

Fig. 3.2 Vessel with an interior carbon pattern characteristic of wet-mode cooking. The wider band of carbon likely resulted from variable water levels

The three seed jars from Sivu'ovi provide the best evidence for cooking. These vessels were found in a covered storage pit and there is no evidence that they were affected by postuse burning. One of the small seed jars (WACC 5918) demonstrates the classic carbon pattern associated with boiling food. The exterior base is slightly oxidized, which is created by having an intense fire under a pot that is raised on rocks or some type of support (Fig. 3.3). The lower third of the exterior wall has a heavy patch of soot, which gradually fades above the midsection toward the rim. The interior of this vessel has the band of carbon that forms in pots used to boil food (Fig. 3.4). A gray carbon patch on the interior base could have been created if most of the moisture had been removed from the vessel in the last stages of cooking.

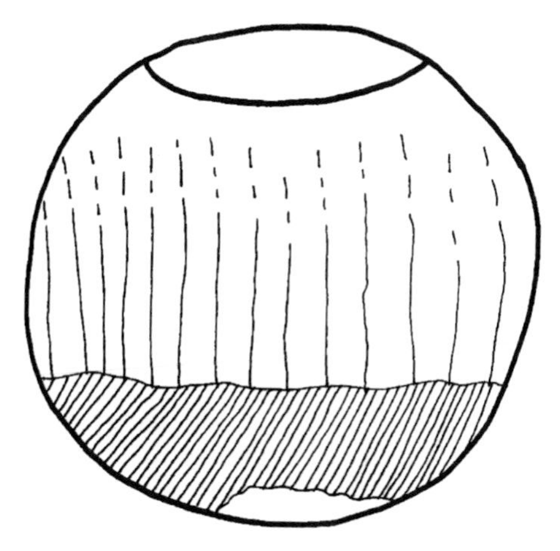

Fig. 3.3 Exterior of a vessel that was used over fire

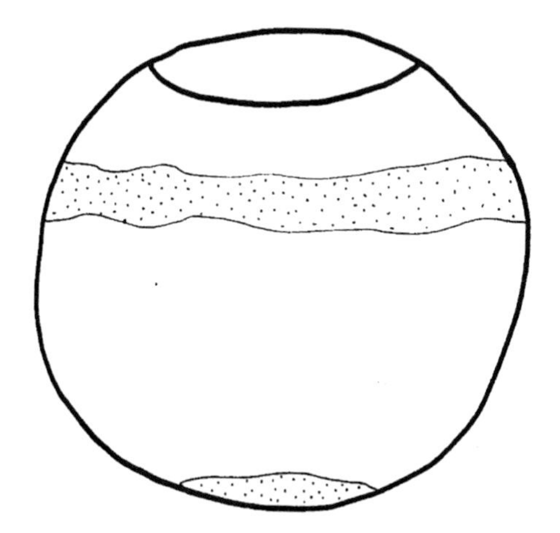

Fig. 3.4 Interior of a vessel used to heat food in the wet mode

The second vessel from Sivu'ovi (WACC 9155) also has clear evidence of use over a fire (Fig. 3.5). This vessel, however, has an interior carbonization pattern that suggests that water was absent during at least some time during most cooking episodes. Water was either removed at the last stage of boiling or food was cooked in the pot in the absence of water.

The largest of the seed jars (WACC 9156) has a similar soot–carbon pattern. The exterior is sooted and the interior has a carbon patch below the midsection, which is caused by heating in the absence of moisture. For food to char it must reach at least 300°C. This can only occur when water is removed from the vessel because temperature in the food below the water line will not exceed 100°C.

This large seed jar also has a heavily abraded interior, which was also observed on nine of the Obelisk seed jars from the Prayer Rock Caves. Only one of these abraded Obelisk Gray pots had evidence of use over a fire, four were not used over a fire, and four were indeterminate. The most likely cause of the abrasion is fermentation. Abrasion by mechanical contact, such as with a scoop or ladle, was ruled out because of the pattern of attrition. In most of the pots with interior abrasion, the entire interior surface was removed, and in other cases the abrasion patch stops abruptly and follows a relatively straight line around the vessel diameter several centimeters below the rim. Such a pattern is more likely caused by the chemical erosion of the interior surface by its liquid contents (Arthur 2003; Hally 1983:19). In low-fired pottery, contents with the opposite pH of the clay can break down the clay structure (Patrick Mc-Govern, personal communication). Thus an acidic ceramic could be broken down by contents with a basic pH, such as the alkaline soaking of maize, and a ceramic with a basic pH can be eroded by acidic solutions. The latter could be caused by the fermentation of some fruits or other highly acidic food. The exact nature of this process, however, is unknown and requires further experimentation.

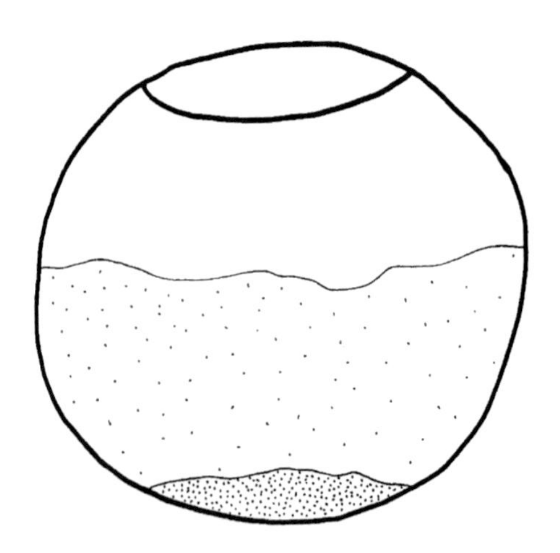

Fig. 3.5 Interior of a vessel used to heat food in the dry mode

Implications

The correlation between seed jar design and function suggests that the vessels could perform well as cooking, storage, or food-processing vessels. The use-alteration analysis demonstrates that the users of this pottery took advantage of their vessel's multifunctionality. There is evidence that some of the vessels were used for cooking (in both the dry and wet modes) and others were not, although the exact function of the noncooking vessels is not known. The heavy interior abrasion on some vessels suggests a chemical erosion most likely caused by fermentation. Organic residue analysis could shed light on what these pots contained. The use-alteration analysis also demonstrated that the vessel users cooked their food in two modes: heating with water and heating without water. The latter can be caused by either cooking dry food (roasting), reheating previously boiled foods, or by boiling something until all or most of the water has been removed. Gruel or stew cooking are cases where enough water could be removed from the contents to cause interior carbon deposits, either as part of the cooking process or by accident.

Southwestern Pottery Origins Revisited

Although the data presented here are just the first step toward understanding the use of early brown ware, we think that they are nonetheless revealing. The earliest pottery on the Colorado Plateau was made by semisedentary pithouse dwellers who began to rely more heavily on maize and other domesticated cultigens (Crown and Wills 1995). They used the multifunctional sturdy seed jars to boil, cook gruel, or reheat a food in the absence of water, for storage, and the fermentation of a liquid that caused the erosion of interior surfaces. Out of the 74 Obelisk Gray vessels from the Prayer Rock Caves only 2 were bowls and 2 were pitchers. One prediction of the Hayden (1995) model is that the earliest pottery would have been dominated by forms used for serving. This expectation is not met at this site because only 6% of the Obelisk Gray vessels were designed for serving. The data presented here agree with the characterization by Crown and Wills (1995) of the context for the adoption of pottery in the Colorado Plateau.

What appears to be happening on the plateau is that the adoption of pottery is a family-by-family decision. The evidence for the brown ware pottery, though widespread, is very scattered. It is likely that between AD 200 and AD 400 there were families that made and used pottery living next to people who did not adopt this technology. The range of early brown ware technological variability also suggests that individuals may have been copying a design (i.e., a seed jar form with sand temper and a roughly polished exterior) but attempting to make it with local resources. Each new potter had to struggle to replicate this design with their own unique local resources.

We do not yet have any direct evidence to infer what was cooked or processed in these pots. Although corn can be processed in new ways with cooking pots, you certainly can effectively prepare corn without ceramic pots, as had been done for centuries. But as Crown and Wills (1995) point out, new variants of maize are also appearing at this time that may have prompted different ways of processing in vessels. Thus, the adoption of pottery could more easily be explained using Brown's model in which people had a greater demand for vessels to store food, soak maize, or store water, but they could not meet the demand with baskets, skins, or some other nonpottery container. Brown's model, however, implies that vessels were not used to solve a particular processing problem. Although we in general agreement with this, we believe that we do not yet have enough evidence for the Southwest to suggest that pots were not used to solve a particular processing need – the boiling of beans.

Beans are the second important cultigen in the great corn, beans, and squash combination that came to dominate the entire Southwest as well as Central and South America. Beans can be soaked and ground into a meal, but by far the most common method to cook beans worldwide is by boiling. The cooking of beans, however, can often take from 2 to 3 h. Long-term simmering of this sort would be tedious with the prepottery cooking technologies. The one great advantage of ceramic pots is their ability to boil foods for long periods with little monitoring. Another advantage of boiling beans instead of some other form of processing is that it reduces the levels of oligosaccharides, the substances that cause flatulence and in some cases extreme abdominal cramping (Stahl 1989:182). Although there is a humorous side to this, it certainly may explain the fact that the most common method of bean preparation is boiling. Intestinal discomfort may in fact play a role in the adoption of pottery on the Colorado Plateau. Certainly, the key to solving this riddle is to further explore how these vessels were used (Longacre 1995:279). Subsequent testing should focus on identifying the organic residues in the early brown ware pottery.

Chapter 4
Smudge Pits and Hide Smoking[1]

From the American Southwest, famous for its pottery, we move to the shores of Lake Superior where the performance-based approach is used instead to explore the function of pit features. These features, given wide notoriety by Binford (1967) in his New Archaeology-type analysis employing analogical reasoning, played an important role in the contact period occupation of Grand Island's Lake Superior shoreline.

Grand Island

Grand Island is located just off the shore of Lake Superior near the present-day town of Munising (Fig. 4.1). The island, the largest on the south shore of Lake Superior, has 35 miles of shoreline and is roughly 7 miles long and 3 miles wide, and covers about 13,600 acres (Roberts 1991:26). The island has two interior lakes, one of which (Echo Lake) is quite large, about a mile in length and a half mile in width. The north side of the island is dominated by sandstone cliffs that are similar to the Pictured Rocks National Lakeshore located on the mainland just east of the island. The southern shore, however, consists of shallow sand or pebble beaches that are protected from the lake's wind and high waves. There is evidence that people have taken advantage of the island's diverse resources from the Archaic Period to the present (Dunham and Anderton 1999).

Not only does the island provide a variety of flora and fauna but there is a historically used sugar maple groove, and the shallows off the south shore are one of the most productive fisheries in this part of Lake Superior. Moreover, the protected bay between the mainland and the island is easily commutable by small boat except in extreme conditions. It is no surprise, therefore, that the earliest Euro-American settlers chose this place for a homestead and trading post, which had been the location for Native Americans for thousands of years (Dunham and Branstner 1995).

[1] This chapter is cowritten by John G. Franzen and Eric C. Drake

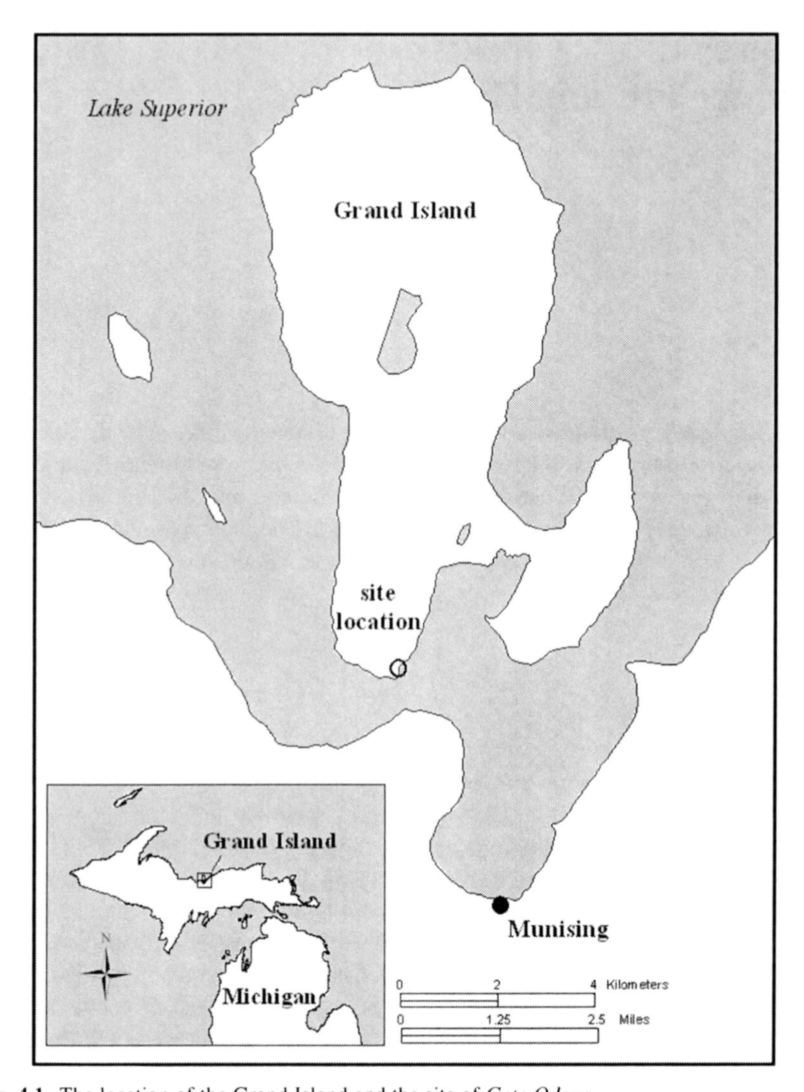

Fig. 4.1 The location of the Grand Island and the site of *Gete Odena*

Gete Odena: Historic Accounts

From the red deer's hide Nokomis
Made a cloak for Hiawatha,
From the red deer's flesh Nokomis
Made a banquet in his honour.
All the village came and feasted,
All the guests praised Hiawatha
(Longfellow 2000:25)

Longfellow's famous book-length poem, *The Song of the Hiawatha*, was based on Ojibwe (*Anishinabeg*) lore collected by Henry Rowe Schoolcraft during his two decades as Indian Agent stationed at Sault Ste. Marie, the important community during the early historic era located in the St. Mary's rapids approximately 140 miles east of Grand Island. From 1822 to 1841, Schoolcraft collected oral histories from the Native Americans that passed in and out of the active trading post. This work was facilitated in large part by his marriage to Jane Johnston, daughter of John Johnstone, one of the most active traders on Lake Superior. Jane's mother was Ojibwe, *Ozhow-Guscodoy-Wayquay* (Woman-of-the-Green-Valley), which gave Schoolcraft easier access to the local Native Americans and permitted him to collect countless stories about their customs, ceremonies, music, and history (Mason 1997).

Schoolcraft's introduction to Lake Superior came about in 1820 when he was asked to join the Cass Expedition. Lewis Cass was then Governor and Superintendent of Indian Affairs of the Michigan Territory, and he organized a trip to explore the southern shore of Lake Superior. Schoolcraft was hired as the geologist and mineralogist and he wrote and published *The Narrative Journal* the following year (Schoolcraft 1821; Williams 1992). This is a detailed account of not only the landscape and mineral resources but of the Native Americans whom they encountered. On 18 June 1820, the group left Sault Ste. Marie following the south shore of Lake Superior and by 21 June they had reached what is now the Pictured Rocks National Lakeshore, which consists of remarkably sculpted 50- to 200-ft sandstone bluffs that rise up from the lake. At about the terminus of Pictured Rocks as you travel west is Grand Island, where the group spent the night.

Here they camped "in a large, deep, and beautiful bay, completely land-locked" (Williams 1992:109,415–416). This is certainly what we call today "Murray Bay," and their camp was either at or very near the location of our excavation project. Schoolcraft goes on to report that "Here we found a village of Chippeway Indians, who, as soon as we landed, came from their lodges to bid us welcome" (Williams 1992:109). That night the camp was the location for dancing, singing, and storytelling. One of the stories was that of the now infamous 13 warriors who traveled to battle with the Sioux after the Grand Island band had been accused of not participating fully in the frequent skirmishes between the two tribes. According to several accounts (see Williams 1992), the Grand Island group engaged the Sioux against all odds and was determined to fight till their death. As they did not want their courage to go unreported, they had the youngest warrior watch the battle from a hidden location where he could witness the event and then report it to their people. The Schoolcraft party was told this story by the young surviving warrior. So impressed was the group that Doty, also a member of the party, published an account of the exploit, "Tale of the Thirteen Chippewas," in the *Detroit Gazette* the following year (Williams 1992:445–446 reproduces the entire *Gazette* article). Although not recorded by any member of the Schoolcraft party, according to local legend the young warrior was "Powers of the Air" who is believed to be represented in a stone carving located on the mainland just 10 miles west of the island. Loren Graham, current occupant of Grand Island's North Light and island historian, has written a popular book that suggests, based on collected oral histories, that the so-called

"Face on the Rock" was made by a member of Schoolcraft's group during a short layover (Graham 1995). Although neither Schoolcraft nor any other member of the party mentions the carving, which is still visible but now badly eroded, this story is deeply embedded in local oral tradition.

Schoolcraft did not say anything else about Grand Island during this trip, but he did on many occasions in later years, as part of his duties as Indian Agent, report on the Grand Island band living on the island. For example, he reports that in 1822 Grand Island had 46 Native Americans (Schoolcraft 1851:102), in 1836 he noted that the Grand Island band consisted of 62 members, and in 1839 he reports that Grand Island had a total of 59 people (7 men, 8 women, and 44 children) (Schoolcraft 1853; see Roberts 1991:49).

About 6 years after the original Schoolcraft and Cass trip, they visited Grand Island again, and this time their journey was recorded by Thomas McKenney. They traveled the south shore of Lake Superior and camped at about the same spot, McKenney believed, that was used by the original Schoolcraft party (McKenney 1959). He mentions an abandoned Ojibwe camp.

> Near our tent I found the frame of a large lodge, and just back of it, the kind of frame on which the Indians dry their fish. It is built over a square hole in the ground, of about six feet by three, where the fire is built. Near the lodge was a pole of a about thirty feet high. At its top hung some badges of the superstition of these people. It was an offering for the sick! From those offerings, we inferred a child had been the subject of their anxieties. Near the top of the pole is a small cap, suspended by a small string – to which is attached, also, a strip of fur. Below these is a little child's covering, not more than ten inches by twelve, with no sleeves, with a feather from the wing of a hawk suspended from near the shoulder-straps. Below, there is a piece of red and white ribband, and ten feet below all, hangs a small hoop, tied round with wattap, which confines to it a parcel of white feathers. (McKenney 1959:362)

Gilman (1836:55) visited the same island location in the fall of 1835 and "found ourselves in the midst of a deserted Indian village." He reports finding the villagers camped on the other side of the island.

A number of other individuals traveled the south shore during this period and many make note of Grand Island (see Castle 1987; Roberts 1991). These accounts and the ones noted earlier, though sketchy, tell us several important things about the native groups on the southwest shore of the island during the period from 1820 to 1840. First, it was a relatively small group. The most accurate estimates were likely made by Schoolcraft and the numbers ranged from 46 to 59 people. Second, many of visitors to the island report the village as "recently abandoned," sometimes with still standing structures. This is in agreement with the notion that the historic and prehistoric groups in the region had a flexible settlement pattern (see Martin 1989, 1999). The location, at best, would have been occupied during the spring through fall but not necessarily on a regular basis. Some of the travelers came to the island during what would have been the prime time for site occupation, only to find it abandoned. The site was clearly occupied on a seasonal basis and not necessarily each year. Third and finally, the southwestern edge of Murray Bay seems to be the consistent location for the historic Native American settlement, which is confirmed by archaeological evidence (Dunham and Anderton 1999; Dunham and Branstner

1995; Skibo et al. 2004). The documents reviewed by Roberts (1991:52–53) suggest that the historic Ojibwe village was located from the sand bluff, on which the Jopling Cottage was constructed (now owned by the Carlsons), to the low-sandy south end of the island. This is a distance of only about 200 m, and our site is located within this zone.

Gete Odena: Williams Era

Abraham Williams, the first permanent white settler on Lake Superior, built his house on Grand Island beginning in 1840 or 1841 at or immediately next to the same historic Ojibwe settlement. Unfortunately, Williams kept no diary, but the evidence of his 33 years in the island is everywhere as some of his structures still stand today. Much of what we do know about this era comes from the work of Castle (1987) who interviewed, in 1906, the 78-year-old daughter of Abraham Williams. Mrs. Trueman Walker Powell, the former Anna Marie Williams, was 12 years old when she arrived on the island and her words provide a vivid account of the early years and the relationship between her father and the local Ojibwe band. According to Mrs. Powell, their family was invited to live on Grand Island by *Omonomonee*, who was "the last chief that had much authority over this tribe" (Castle 1987:32). We take the name "*Gete Odena*," which means "ancient village," from the Williams' era. The Ojibwe settlement on the island at the time was referred to as the *Gete Odena*, near which Williams built his home.

Our site, which is within 100 m of one of the homes built by Williams, may be the *Gete Odena* as we have a strong Late Woodland occupation at the site, but it was also occupied during the historic period prior to and even after the arrival of Williams. Thus, the descriptions provided by Powell are especially relevant to this discussion. Powell notes that the Ojibwe lived on the island only in the summer and describes their village:

> Saplings were set into the ground at regular intervals and their tops were tied together to make a roof....This framework was covered with square mats which lapped one over the other, and which were made of the long leaves of the "cat-tails" woven on a woof of tough roots. The bark of the basswood tree were also used. These mats were practically indestructible, and possessed the further advantage of being easily removed and set upon another framework...In the center of every lodge was an open fire. Around the sides were the beds, made of furs flung on hemlock boughs. (Castle 1987:36)

Williams was a "man of parts" (Roberts 1991:96). He was a blacksmith, cooper, carpenter, farmer, fisherman, and trader. The Williams family arrived on the island at the end of the most productive fur trade period. The Ojibwe in the region had contact with the French traders beginning in the 1600s followed by the British and then Americans. By the early 1800s, the prized beaver had started to become scarce and the major fur trading activity moved west (Bishop 1974:11–12). There continued, however, active trade in other fur-bearing animals such as muskrat, marten, moose, and deer. Williams involved himself immediately in this trade and was successful in taking business away from the American Fur Trading outposts on Lake Superior.[2] Williams also obtained fish from the locals, which he put into barrels he made on Grand Island. In the 1850s, it was reported that Williams was producing each season several hundred barrels of fish, each holding about 200 pounds (Roberts 1991:102).

Besides trading with the local Ojibwe, Williams also hired them for various activities. Although there is little direct evidence that Williams employed the Ojibwe, one only needs to look at his accomplishments to envision that he must have had a group of locals employed most times. Besides building numerous structures on Grand Island, he also built a sawmill on the mainland, worked as a carpenter in the newly founded city of Marquette, built hundreds of barrels each winter, farmed, operated a blacksmith shop, supplied firewood for the steam ships on Lake Superior, and operated a brisk trade with the Ojibwe in furs, fish, and various other items. Although his wife and children were working on these projects, there is some evidence that Native Americans were hired as well. Brotherton (1944:198–203) visited the island in 1853 on a steamship and notes, "Indians in the employ of Williams began loading dry hardwood cut in four foot lengths as fuel for out steam boilers." Clearly, Williams and the local Ojibwe developed a symbiotic relationship in which the locals provided Williams with furs, fish, and labor, and Williams turned a handsome profit from the transactions as well as providing the Ojibwe with the trade goods they desired. This was such an important relationship during this period that once Williams left the island the Native Americans did as well. Williams died in 1873, and the 1880 census lists no Native American on the island (Roberts 1991:62). The Native Americans, as well as Williams' descendants who stayed in the area, moved to the mainland.

Gete Odena: Smudge Pits

Several dozen features were exposed during two excavation seasons, and six of them were unique pit features that were consistent in both morphology and content. The exact forms of two of the pit features are not known because they were truncated by later disturbance. The content of these features, however, was identical to the other four and thus we think that they functioned in the same manner.

The pits had a mean maximum width of 36 cm and a mean maximum depth of 46 cm. As seen in Fig. 4.2, three of the pits are slightly bulbous in profile. The base of each pit has a layer of charred, half-burned fuel that was in such a good state of preservation that in some cases pine cones were still intact and needles could be identified. Each pit was filled with sandy, mottled soil (light brown through dark brown in color) with flecks of charcoal.

---

[2] The American Fur Company Traders on Lake Superior complained to their superiors that Williams was trading whiskey to the Ojibwe, which was in violation of the 1842 Treaty of La Pointe (Roberts 1991: 57)

Feature 3 has a slightly different shape and fill sediment. The sides of this pit are straighter and the fill, though still quite mottled in appearance, has a much darker micro-strata. We suggest that this pit was reused as least once and possibly several times. When initially dug, it may have had the same bulbous shape as the other three pits, but after one or more reuses, the sides gradually became straight. The bulbous shape, with a narrowing at the midsection, would not hold up long in the sandy soil as it dried. Collapsing walls were an ongoing problem during excavation,

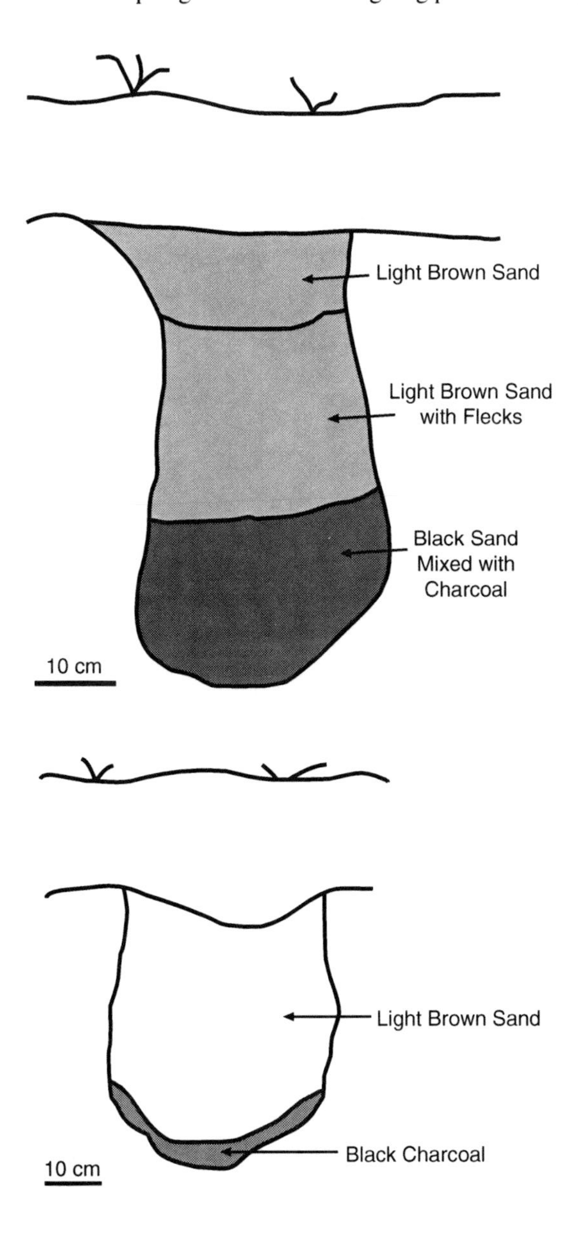

(Continued)

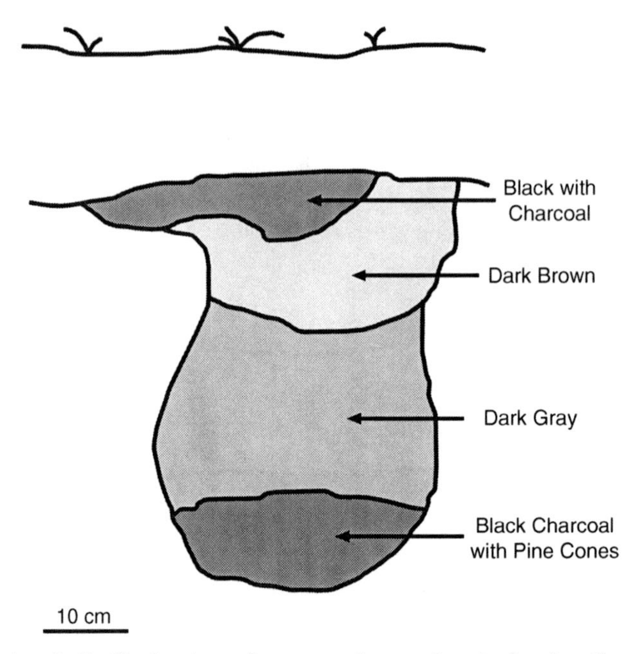

Fig. 4.2 (continued) Profile drawings of representative smudge pits found at *Gete Odena*, each showing charred material at the base

which suggests to us that the shape of the bulbous pits could not be maintained without being filled in soon after use. The walls of such a pit would tend to become straighter through time and with repeated contact with arms and hands during preparation of the smudge fire. Supporting this multiple use argument is the denser, thicker, carbonized layer at the base of Feature 3.

Performance-Based Analysis

Binford (1967) elevated similar pit features to considerable fame as he used them to demonstrate analogical reasoning with ethnographic sources. He excavated 15 of these features at a Mississippian site excavated as part of the Carlyle Reservoir project in southern Illinois (Binford et al. 1964). The pits had a mean width of about 27 cm and a depth of 30 cm. The fill of these pits is remarkably similar to what was found at *Gete Odena*, the base of the features had charred, and carbonized organic matter and the remainder of the pit fill had a grayish loam soil that demonstrates "intentional filling of the pit contents" (Binford 1967:38). One notable difference between the Mississippian pits and those found at our site is that the former often contained remnants of corn cobs as part of the carbonized fill. The botanical analysis from *Gete Odena*, to date, has not recovered any evidence of corn, which would preclude its use as smudge material. The pits at our site are also slightly larger than those recorded by Binford. This could be the result of slightly different functions, but it also may be because the sandy soil at *Gete Odena* lends itself to easy digging even with one's bare hands. In the original archaeological report, Binford et al. (1964:17) suggested that the features were smudge pits that could have been used to keep away mosquitoes.

Not satisfied with this explanation, Binford in his later paper on analogy (Binford 1967) did an ethnographic search and found 13 groups, ranging form the Plains to the Southeast and Great Lakes, with a reported use of similar features. In all cases, the pits were used to smoke hides. Binford then concluded, based on the similarity in form, size, and content between the prehistoric and ethnographic cases, that the 15 features from his site were used to smoke hides. Spector (1975) identified almost identical features at the late-eighteenth-century Winnebago site in southeastern Wisconsin. Charred corn cobs were also found at the base of each of the pits.

Munson (1969) suggests that Binford's functional interpretation of the smudge pits was too narrow. In fact, he cites several ethnographic cases in which similar pits were used to smudge pottery, not smoke hides. He does not suggest that Binford's smudge pit argument was incorrect, only that it may be too narrow. Binford (1972:53–58) counters Munson's argument by agreeing that it is indeed possible to smudge pots using pits (and there is some ethnographic cases to back this up), but the most common way to smudge pottery does not involve using a pit. He then argues that the strongest argument, based on analogical reasoning, can still be made for hide smoking, instead of pottery smudging or any other use of smudge pits.

The Binford and Munson debate is interesting because it demonstrates how the New Archaeology began to grapple with the use of analogy and hypothesis testing in their quest to reconstruct prehistoric activity. We argue, however, that while ethnographic analogy does indeed play a critical role in understanding the function of features or artifacts, an equally important step is a performance-based analysis of the features themselves.

Performance Analysis of the Features

As noted previously, performance characteristics are the characteristics an artifact or feature must possess in order to perform its function. Performance characteristics and the associated technical choices can be inferred by isolating the attributes of the feature or artifact. This is a theory of artifact or feature design that answers the question, "Why was this feature made in this way?" Isolating the technical choices along the feature's entire life history provides the clues to understand performance characteristics. In this case, we can infer these technical choices by the stratigraphic information, pit contents, and other contextual information. There are two performance characteristics that we can infer from the technical properties of the pits: oxygen deficiency and ease manufacture.

Oxygen Deficiency

The life history of these features begins with the need for a pit. Individuals have lots of options at this point that are governed by the function of the pit, the soil type (rocky, sandy, etc.), the time they want to invest in the project, and the available tools for excavation. The formal properties of the pits can be used to infer the technical choices made by those who dug and used the pit. Each pit has a relatively narrow mouth that would mean that only one arm could be used for excavation by hand or with the help of a simple scoop. The mean maximum depth is about 50 cm, which is about as deep as a pit can be dug by hand. A smoky fire could be built on the ground or in a shallow, basin-type pit, but they chose to dig a relatively deep, narrow pit. A fire at the base of such a pit would create an oxygen-deficient environment – a smoky, smoldering, flameless fire. Hilger (1992:132) observed smoking hides on the La Pointe Reservation in which a metal bucket was used instead of a pit. The bucket functioned adequately for smoking the hides but it had to be monitored carefully so that the flames would not erupt and burn the hide. Careful monitoring of this type would not be necessary if smudging were done in the deep, narrow pits at *Gete Odena*.

The content of the pits also provides important clues to their choice of fuels. Dry wood, grass, or other dry fuel would be readily available, but they chose, in this case, pine bows and cones. Ritzenthaler and Ritzenthaler (1983:82) note that the ideal smudge fuel would be "rotten pine or poplar, or in some cases, Norway pine cones," and Hilger (1992:132) observed white-pine and Norway pine cones being used. At the base of each of the *Gete Odena* pits were charred cones and pine twigs that still maintained their structural integrity. In an oxygen-rich environment, even green bows and cones will combust.

Ease of Manufacture/Expediency of Manufacture

The sandy soil matrix at the site would make it possible for a person to dig a pit of this type in a matter of minutes. What is more, the unstable sandy soil would not permit these pits to be left open for long without caving. The stratagraphic evidence suggests that the pits were dug, used, and filled within a short period of time, likely in the same day. As mentioned earlier, the bulbous shape of the pits could not be maintained if the pit was left open for any length of time. In addition, the great quantity of unburnt matter at the base of the pits suggests that pits may have been filled immediately after use while the smudge was still smoldering. Ritzenthaler and Ritzenthaler (1983:82) report that coloring of hide over a smudge pit would take only about 15–20 min per side. With the exception of Feature 3, which had the straight sides and deeper charred layer, it is quite likely that the pits were used just once. When one or more hides were ready to be smoked (they could be sewn together and smoked at the same time), a pit was dug, pine cones and other smudge material was put in the base, and a few burning coals were added to start the smudge. After roughly half an hour, the hide was removed and the dirt was tossed back into the pit to extinguish the fire and cover the hole.

The above performance-based analysis suggests only that the pits were well suited to create a smudge fire on an as-needed basis. They are easy to make and easier to fill up when done. The question remains, however, What were these pits used for? To answer this we turn to contextual clues, faunal data, and the ethnographic record.

There is no evidence among the Ojibwe that they smudged the interior of their vessels. Moreover, the single C14 date run on a cone from the pit along with other contextual information suggests that the pits were used somewhere between 1810 and 1850. Although this was prior to or immediately after the arrival of Abraham Williams, there is no evidence that traditional handmade pottery was made at this time. The mouths of the pits at our site are also too big for the standard Late Woodland vessel. Munson (1969) found some ethnographic support for smudging pots over a pit, but this is clearly the exception worldwide. The typical pattern is to remove a hot pot from the fire and place it directly over combustible material.

We do, however, have ethnographic cases among the Ojibwe for smoking hides using very similar features (Buffalohead 1983; Densmore 1979; Hilger 1992; Johnstone 1990; Ritzenthaler and Ritzenthaler 1983). Figure 4.3 shows photos taken by Densmore (1979) in the early twentieth century.

> If several hides were to be smoked, they were sewn together in such a manner that they formed....(a) conical shape.....A hole was dug about 18 inches in diameter and 9 inches deep. Over this a framework was constructed that resembled a small tipi frame. The hide was suspended above the framework and drawn down over it...A fire had previously been

BUREAU OF AMERICAN ETHNOLOGY BULLETIN 86 PLATE 75

SMOKING DEER HIDE

Fig. 4.3 Photos of an Ojibwe woman smoking a deer hide (Bureau of American Ethnology, Bulletin 86, Plate 75)

made in the hole, Zozed using dry corncobs for the purpose. This fire smolders slowly, the smoke giving to the hide a golden yellow color. (Densmore 1979:164–165, Plate 75)

Other Supporting Evidence

Supporting the hide-smoking argument is the large number of hide-bearing animal remains identified during the faunal analysis, which includes beaver, deer, muskrat, wolf, marten, otter, and moose (Skibo et al. 2004). Although we cannot make a direct correlation between the bones recovered and the construction of the smudge pits, the patterns in the faunal material are striking and may be of some note. During the 2001 and 2002 seasons, a total of over 1,400 pieces of bone were recovered (see Skibo et al. 2004 for a complete discussion). A total of 89.6% of the bone came from mammals. The vast majority of the mammal bone was too small and fragmentary for species identification. Beaver was the most common animal bone identified (42 pieces), though the bones represented just 3 MNI. The most surprising result was the small number of fish bones recovered. Just two sturgeon, two whitefish, two Walleye, and one Channel catfish were identified, and there were only 30 total fish bones identified. For comparison, at the Juntunen site, 85% of the recovered bone came from fish (McPherron 1967), whereas fish make up just 2% of our collection. This is surprising because the site is located at a classic location for a Great Lake's fishing village, and just off shore of the island is one of the most productive fisheries in the south shore of the lake. The overall distribution of bone species along with the presence of the smudge pits, discussed earlier, certainly supports the notion that this section of the site was involved in hide processing during historic period.

Why Smoke Hides?

Ritzenthaler (1949; see also Ritzenthaler and Ritzenthaler 1983) discusses what he calls the "tanning process." After the skin has been removed from the deer it can be sold as a "green" hide, or the hide can be processed further. When Ritzenthaler (1949) recorded the process, a tanned hide could get up to double the price of a green hide. The men shoot the deer and remove the hide but women did the remainder of the process, which first involves removing the hair. This was done on a "beaming" pole, which is simply a smooth peeled log. The hide is placed on the pole and the hair is removed with a scraper, which was a cylindrical piece of wood imbedded with a blunt table knife (Ritzenthaler 1949). After the hair is removed the hide is soaked in a solution of warm water and dried deer brains, which had been initially prepared by boiling in a frying pan. If brains are not available then egg white is used. Holes are then cut into the edges of the hide and it is attached to stretching frame. The hide is stretched and then allowed to dry on the rack. The smoking process involves sewing the edges of the hide together to form a "cylindrical bag" (Ritzenthaler 1949:11–12). The group observed by Ritzenthaler also sewed a piece of cloth onto the bottom of the hide that was then attached to the metal smudge bucket. This additional cloth was needed to keep the hide from burning, which would be unnecessary if a smudge pit was used. The hide is then smoked for about 15 min per side, which is a process called "*sowagige'akwans*" (Hilger 1992:132). Ritzenthaler (1949:13) also notes that the summer hides are preferred for this process because they are thinner and tougher. The best months are July and August.

Historical evidence provides a context for the possible abundance of smudge pit features in early-nineteenth-century Ojibwe settlements. By the 1830s, beaver and other fine furs such as marten, fisher, and otter were relatively scarce, and muskrat and deer dominated the inventories of traders south of Lake Superior (Gilman 1974:18). These inventories often specify that deerskins received from Native Americans were processed (smoked). Although deerskins were much less valuable than less common species, the American Fur Company would still purchase deerskin at prices that precluded a profit just to keep them away from competitors and maintain trading relationships (Peake 1954:246–247). An early-nineteenth-century XY Company trader in northern Wisconsin mentions trading for "dressed deerskins" (presumably stretched and smoked), as do western traders of the same period (Curot 1911:412; Work 1914:269).

Smoking of deerskin was essential when used for making moccasins because this enabled them to remain soft despite repeated wetting and drying. In fact, based on a survey of ethnographic and ethnohistoric literature, Richards (1966) concludes that "smoking was more important in northern (wetter and colder) regions and moccasin hides were the most likely recipients." George Catlin (1985) observed the process in the northern plains and notes that "heated smoke; and some chemical process or other, which I do not understand, the skins thus acquire a quality which enables them, after being ever so many times wet, to dry soft and pliant as they were before, which secret I have never yet seen practiced in my own country." The influx of nonnatives who needed footgear into the area, as well as the emerging broader market for Indian craft items, may have actually increased the incidence of deer hide procurement and processing during the early nineteenth century. Shoes, boots, and imported leather are absent in surveyed trade good inventories from this period, which are dominated by textiles, so moccasin leather would have been needed for both home use and trade (Michigan Pioneer Historical Society 1985; Johnston 1822; Kinzie 1932; Thwaites 1910). The combined depletion of beaver populations and possible increased demand for tanned deerskin could explain why smudge pits are conspicuous at *Gete Odena*, as well as contemporary sites such as the Cater Site (Beld 2001). Kinzie (1932:13–14) lists various items brought in for trade in northern Wisconsin c. 1830, and includes smoked deerskins, moccasins, and hunting pouches. She also describes the outfitting of a typical *voyageur* as including "one or two smoked deerskins for moccasins" (Kinzie 1932:229). It is notable that traders in the Southeastern USA during this same period commonly obtained deerskin for export. The mid-eighteenth-century trade between the French and Creek Indians in what is now Alabama, according to Waselkov (1992:37), was dominated by the trade of deerskins. Up to 60,000 deerskins were shipped to France from the "French Louisiane" in 1860. Three types of skins were traded. The first, referred to as "dressed skins" discussed earlier, were "stretched, scraped on both sides, treated with deer brains, and finally smoked" (Waselkov 1992:37). Skins with the hair intact were used by French tanners to produce "Moroccan grain leather," whereas skins that had been scraped but not stretched or smoked were made into parchment for binding books.

Conclusion

Gete Odena was occupied seasonally from for over 600 years. The function and use of the site no doubt varied considerably over these years, but this chapter focused on the late-eighteenth and early-nineteenth-century occupation. During this period, a number of smudge pits were constructed presumably used to smoke hides (likely deer). A performance-based analysis of the pits suggests that they were likely single-use features designed to create a smoking fire in an oxygen-deficient environment. The dominance of large mammal bones from the site, instead of fish, also suggests that this site was used in part for the processing of large game. Historic and ethnohistoric sources also suggest how these features were used and why the smoking of deer hides became increasingly important during this period. Besides the fact that other more valuable fur-bearing animals were scarce during this time, there was an increased demand among immigrants to the area for smoked deer hides that could be used for making moccasins, coats, and leggings.

Chapter 5
The Devil is in the Details

Since the early 1990s, archaeologists have shown a heightened interest in explaining technological change. Indeed, this general research goal is now supported by archaeologists of every theoretical persuasion (e.g., Bleed 2001b; Dobres 2000; Dobres and Hoffman 1999; Fitzhugh 2001; Gordon and Killick 1993; Gould 2001; Hayden 1998; Hughes 1998; Kelly 2000; Killick 2004; Kuhn and Sarther 2000; Lemonnier 2000b; Mithen 1998; Neff 1992; O'Brien et al. 1994; Rice 1999; Roux 2003; Sassaman 1993; Schiffer 1992, 2001a; Schiffer et al. 2001; Shott 1997; Sillar and Tite 2000; Skibo 1994; Stark 2003). Moreover, most archaeologists agree that technologies are context dependent, their form and prevalence contingent upon local, historically constituted conditions. Thus, specific explanations are tied to a given group in time and space and are richly supplied with relevant particulars of the societal context. On the basis of these contingencies, the archaeologist fashions an empirically grounded narrative that accounts for a given technological change.

The provision of historical narratives is not the exclusive aim of technological studies because archaeologists also craft crosscutting theories and models. This strategy is pursued when researchers ask, and seek answers to, *general* questions – those lacking time–space parameters. Although archaeologists have offered generalizations about processes of technological change, such as adoption or consumption (e.g., Spencer-Wood 1987; see also references in Schiffer 2001b; Schiffer et al. 2001; van der Leeuw and Torrence 1989), invention processes have been woefully undertheorized (Fitzhugh 2001).

If we aim to achieve a comprehensive understanding of technological change, then our corpus of principles must come to include generalizations about the sources of material novelty. After all, invention is a commonplace human behavior, and so its study offers an opportunity to fashion principles of great generality. Fortunately, recent efforts suggest that at least some invention processes are patterned and can be described by models and theories (e.g., Fitzhugh 2001; Hayden 1998; Schiffer 1993, 1996, 2002).

Given that myriad activities can generate material novelty, the first task is to identify behaviorally based kinds of invention processes. Each kind of process is operative in a specific "behavioral context" (LaMotta and Schiffer 2001; Walker et al. 1995), an analytic unit defined by shared "characteristics among seemingly dissimilar – often culturally diverse – empirical phenomena" (Schiffer 1996:651). The second

J. Skibo and M.B. Schiffer, *People and Things: A Behavioral Approach to Material Culture*

task is to devise the theory or model that best accounts *in general terms* for the operation of each kind of process. Our expectation is that by defining and studying varied behavioral contexts, we can create a family of generalizations that encompass diverse invention processes. Given these intellectual resources, the archaeologist could not only furnish a contextualized narrative of a given invention, but could also invoke the appropriate model or theory, which would implicate the relevant nonunique factors that tie the case to many others.

This chapter focuses on the kind of invention processes that arise in the behavioral context of "complex technological systems" (CTS). I define CTS as any technology that consists of a set of interacting artifacts; interactions among these artifacts – and people and sometimes environmental phenomena – enable that system to function. Because the archaeologist has wide latitude in interpreting the terms of this definition and because technological complexity is ostensibly a continuum (e.g., Oswalt 1976), the determination of whether a specific technology constitutes a CTS is necessarily driven by the archaeologist's research problem. Given the flexibility of this definition, one can expect to discern CTSs in diverse – even small-scale – societies (see "Operationalizing the Cascade Model on Archaeological Cases").

For handling CTS-related invention processes, a "cascade" model is presented, which is a behavioral adaptation, elaboration, and generalization of Thomas P. Hughes' (1983) model of "reverse salients." According to Hughes, during the development of a complex sociotechnical system, like an electric power network, certain components lag and present critical problems, such as generators of insufficient capacity to meet demand and power poles vulnerable to lighting strikes. If the system is to become functional, then such problems must be solved – usually through invention. Hughes' model, especially the notion of lag and the construct of reverse salient, derives from a military metaphor that implies grand, if not grandiose, development campaigns. On the other hand, the cascade model is expressed in terms that appear to fit a wider range of CTSs, including those that might be present in small-scale societies. Also, the cascade model seems well suited to explain the serial spurts of inventive activities that accompany a developing CTS (see also Gould 2001).

In a nutshell, the cascade model posits that, during a CTS's development, emergent performance problems – recognized by people as shortcomings in that technology's constituent interactions – stimulate sequential spurts of invention. As adopted inventions solve one problem, people encounter new and often unanticipated performance problems, which stimulate more inventive spurts, and so on. The result is a series of "invention cascades." A distinctive feature of the model, which promotes its generality, is the premise that processes in a CTS's life history are the immediate contexts in which performance problems emerge and stimulate invention cascades. Thus, life-history processes are suitable analytical units for investigating invention processes in CTSs.

It is important to emphasize that the cascade model does not explain how or why the development of a CTS is initiated; rather, it accounts for the spurts of inventive activities that transpire during the course of development.

This chapter has five major sections: (1) general considerations concerning CTS-related invention processes, (2) elaboration of the cascade model, (3) illustration of the model with the development of the nineteenth-century electromagnetic telegraph, (4) discussion of the model's applicability to small-scale societies, and (5) enumeration of the model's broadest implications for studying technological change.

General Considerations

We begin by presenting definitions tailored to the cascade model. "Invention" is the activity that creates a novel *technological object* or artifact – that is, a new kind of part, assembly, component, or subsystem. To qualify as "new," a technological object is expected to differ, *in one or more performance characteristics*, from other artifacts in the same society. Clearly, archaeologists can consider only inventions that have been materialized in some form (e.g., drawings, models, full-scale hardware). The term "inventor" designates not an occupational specialization but the person, task group, corporate group, or collective that created the new technological object.

An important premise of the cascade model is that, in a CTS's development, people respond to each performance problem by engaging in inventive activities until one or more of the resultant technological objects contributes to an acceptable solution. Thus, a performance problem usually causes inventors to generate a set of technological variants, from which individuals (and social units at various scales) select for incorporation into other activities. For example, artisans – as manufacturers – elect to replicate only some inventions, which are further winnowed by consumers (on the "replicative success" of artifacts, see Leonard and Jones 1987). Needless to say, when they are subjects of explanation, replication (or manufacture) and consumption (or adoption) require their own models (see Schiffer et al. 2001). Although a source of variation subject to selection, invention processes are far from random, and so are not equivalent to genetic mutation (cf. Fitzhugh 2001; see Schiffer 1996 on "stimulated variation").

Generalizing from historical examples, it is suggested that most inventions – even those that become hardware – are unsuccessful owing to shortcomings in performance characteristics; they are neither replicated by artisans nor adopted by consumers. Successful inventions are evidently a small, and almost certainly unrepresentative, sample of the products of human creativity. If we aspire to construct *general* theories and models, then it behooves us to consider all (knowable) technological objects, successful or not, that result from an invention process. Otherwise, our narratives are apt to consist of presentistic chronicles of only replicated and adopted technological objects.

The relentless variety-generating feature of invention processes has straightforward implications for understanding archaeological variability. Variants that became hardware but were judged unsuitable usually end up being reused or discarded.

In either case, barring deterioration, the remains may be included in archaeological deposits. Thus, the cascade model can help the archaeologist to seek, identify, and explain certain patterns of formal variability that might otherwise elude scrutiny.

In building and illustrating the cascade model, a well-documented case from the historical record is used. As Dethlefsen and Deetz (1966) demonstrated long ago in their studies of New England gravestones, the historical record is fertile ground for cultivating new archaeological method and theory (see also South 1977). Inventions that arose in the course of creating the electromagnetic telegraph are lavishly recorded in a huge technical literature. These writings furnish information about the proximate contexts of invention cascades and on the countless technological objects, successful and unsuccessful, that they begat. On the criterion of sufficient surviving evidence, the electromagnetic telegraph is an ideal case. In drawing inspiration from a CTS in a capitalist-industrial society, I have strived to fashion a model that by virtue of its generality and flexibility is also applicable to small-scale societies (see "Operationalizing the Cascade Model on Archaeological Cases" and "Discussion").

CTSs of great complexity, like the electromagnetic telegraph, have three common concomitants. First, some require a complex social organization, with many people performing specialized, hierarchically related roles, as in automobile factories, churches, and ships. Second, a number of CTSs, such as an electric power grid, road network, and canal irrigation system, exhibit considerable spatial extension. And third, CTSs having complex social organizations and great spatial extent tend to endure for many decades, sometimes centuries. These concomitants are most likely to co-occur in the technologies of complex societies. Indeed, the terms "socio-technical system" (Hughes 1983) and "large technical system" (Joerges 1988), which were formulated by historians to handle certain Western industrial technologies, imply both organizational complexity and spatial extension. However, these are *not* essential features of a CTS, as defined here. CTSs and cascade invention processes can also occur in small-scale societies (see "Operationalizing the Cascade Model on Archaeological Cases").

Another scale issue that enters into the designation of CTSs is that of bounding the unit of study. For example, in investigating the telegraph, we may choose any of the following: (1) one telegraph, (2) one telegraph network, (3) all telegraphs in one nation, or (4) all telegraphs in the world. Because the telegraph developed as a result of inventions made in several nations, by members of an international community of inventors competing for patents, financial support, employment, prestige, and social power, it is justifiable to choose the largest scale – that is, all telegraphs. Nonetheless, the American Morse telegraph, which was eventually adopted throughout the world, serves overwhelmingly in the examples below and effectively illustrates the model.

In the past few decades, students of technology in many disciplines have properly called for greater efforts to show how technologies develop in response to a variety of contextual factors – for example, religious, economic, political, social, and ecological (Adams 1999; Arnold 1993; Bijker 1995; Dobres 2000; Dobres and Hoffman 1999; Galison 2003; Hughes 1983; Killick 2004; McGuire and Schiffer

1983; Mom 2004; Mills and Crown 1995; Nelson 1991; Schiffer and Skibo 1987; Schiffer et al. 1994a; Shackel 1996; Skibo and Schiffer 2001; Staudenmaier 1985, 2002). It is widely appreciated that people in virtually any realm of society, from religious leaders to subsistence farmers, can foment the development of new technologies. Moreover, the actual course of development depends greatly on the kinds of social roles and social units available to underwrite the process, such as branches of government, political leaders, stock-issuing corporations, communities, religious congregations, elites, kin groups, aggrandizers (Hayden 1998), sodalities, households, and task groups. Such organizational variation can affect, for example, the resources available to support and reward inventive activities, acceptable values of core performance characteristics, decisions to pursue development, strategies for developing CTSs, and ultimate outcomes (e.g., Galison 2003; Hughes 1983). Although the cascade model itself draws attention mainly to the *proximate* contexts of invention processes, both proximate and distant contextual factors are essential for crafting well-rounded, anthropologically sound narratives of technological change (Fitzhugh 2001). Needless to say, identifying the more distant contextual factors and linking them *rigorously* to specific technological changes is the creative challenge we all face.

It is also important to note that performance problems in a developing CTS are sometimes solved by organizational inventions (Chandler 1977), including new ways to recruit, train, and discipline workers. Such solutions, however, are not within the cascade model's compass. Perhaps archaeologists whose inspiration comes from other theoretical programs, such as social construction (Killick 2004) or agency theory (Dobres 2000; Dobres and Hoffman 1999), can build models for handling all responses to performance problems.

The Cascade Model

A CTS has a life history consisting of a minimal set of processes: creating the prototype, replication or manufacture, use, and maintenance. These processes, however, do *not* comprise a unilinear sequence; some may occur in parallel and others can recur. Depending on the CTS and one's research interests, many more processes can be specified. Thus, to accommodate the telegraph's diverse invention cascades, a large set of processes is delineated (some of which may apply only to CTSs in capitalist-industrial societies). Although the model can be elaborated ad infinitum, a key premise remains invariant: life-history processes, however subdivided, are the proximate contexts of invention cascades. By employing life-history processes as analytic units, one can operationalize the model systematically (see below).

A life-history process consists of interrelated activities, which in turn incorporate one or more technological objects. If the CTS's life history is to have a forward motion – that is, proceeding from activity to activity and from one process to the next – people must judge that the technological objects have reached acceptable values of "core" or "critical" performance characteristics (see Schiffer and Skibo

1997). As behavioral capabilities, performance characteristics can enable any kind of interaction – for example, mechanical, electrical, thermal, or chemical. In addition, many performance characteristics pertain to human senses, such as olfactory, gustatory, tactile, and visual, and facilitate symbolic behavior (Schiffer and Miller 1999a). The effort to achieve acceptable values of critical performance characteristics – whether utilitarian or symbolic – usually provokes a spurt of inventions, which can in turn foster further spurts. Each life history process, consisting of activities and their constituent interactions, is a potential incubator of invention cascades.

The minimal unit of an invention cascade is a flurry of inventions that tend to cluster somewhat in time but not necessarily in space. As variants of a particular kind of technological object, defined on the basis of utilitarian and/or symbolic functions, the inventions usually differ in how well they achieve the critical performance characteristics. These performance differences affect selection processes: many inventions are judged unsuitable and are not replicated; some, though promising, are replicated but only sporadically adopted; and others, regarded as successful, are replicated and adopted widely. In some cases, no suitable variants are invented, which truncates or radically redirects the CTS's development.

Cascades can occur at any scale of technological object, from part to subsystem; in very complex CTSs, one often finds a hierarchy of invention cascades. For example, in the 1890s, when marked interest arose in building automobiles, there was a cascade of prototype vehicles with different motive powers: steam, electricity, gasoline, compressed air, and even springs (Hiscox 1900). Manufacturers quickly selected in favor of gas, steam, and electric. Inventors in turn created countless alternative designs for specific parts, assemblies, and so on for each vehicle type. Among the cascades that arose were inventions for ignition and cooling systems in gasoline automobiles, for batteries and controllers in electrics, and for boilers and condensers in steamers. During the next two decades, the symbolic functions of gasoline and electric cars also stimulated invention cascades in body styles and interior furnishings (Mom 2004; Schiffer et al. 1994a). As in the automobile case, inventors may initially adopt different approaches to achieving the CTS's core performance characteristics, leading to diverse technological objects at many scales. Gould (2001:201) has compared the proliferation of early steamship designs to "adaptive radiations" in biology.

In capitalist-industrial societies especially, CTSs sometimes undergo a succession of invention cascades lasting many decades or even centuries (cf. Mokyr 1990). Indeed, the gasoline automobile in the twentieth century experienced virtually continuous cascades. Major cascades arose, for example, in response to changes in contextual factors, such as fuel costs, road design, and governmental regulations, which affected the criticality of performance characteristics relating to fuel economy, puncture and wear resistance of tires, and permissible quantities of exhaust chemicals. In addition, the adoption of a technological object can alter the performance requirements of other objects with which it interacts, leading to further cascades (on such "disjunctions," see Schiffer 1992, Chap. 4). CTSs in small-scale societies, such as canal irrigation systems, also would have experienced, one would think, more or less continuous invention cascades.

Illustrating the Model: The Electromagnetic Telegraph

This section, which treats the electromagnetic telegraph, serves several purposes beyond illustrating the cascade model. First, it defines along the way the four basic processes (i.e., creating a prototype, replication or manufacture, use, and maintenance) in more detail. Second, it demonstrates how easily the cascade model can be elaborated beyond the four basic processes. Third, this section calls attention to the host of unsuccessful technological objects that an inventive spurt can leave in its wake, which can potentially reach the archaeological record. Fourth, it emphasizes that many kinds of performance characteristics, utilitarian and symbolic, become critical in specific life-history processes. And fifth, it instantiates the behavioral tenet that archaeologists can study people–artifact interactions in any society, without regard to time or space (Reid et al. 1975).

Creating the Prototype

A CTS often begins its life as an idea or vision for a technology that is expected to have certain use-related performance characteristics. In capitalist-industrial societies, these visions have many sources, including existing technologies; previous but unsuccessful attempts to construct a similar CTS; literatures of science, engineering, and popular culture – including science fiction; playfulness of creative people; and "cultural imperatives" (*sensu* Schiffer 1993). Often, the vision arises independently among many individuals. Indeed, in a community of practice, such as electrical experimenters, astronomers, or shipbuilders, ideas for a new CTS may be obvious to its more knowledgeable members. As the telegraph case makes clear, however, the hard work of inventing is in the details, in working out the CTS's numerous "little" inventions that comprise cascades.

Captivated by the vision, inventors strive to make prototypes that exhibit minimal functioning. "Minimal functioning" means the achievement of the CTS's core performance characteristics at a level merely adequate to demonstrate to the inventor (and perhaps kin, friends, or associates) that such a system is technically possible. Constructing a prototype often leads to many invention cascades.

"Telegraph" was already a familiar term in the early nineteenth century, for by then various mechanical–optical telegraphs, such as semaphores, had been operating in France, Germany, and England (Shaffner 1859). Indeed, all of France had been knit into a single, government-controlled network centered on Paris (Beauchamp 2001). Limited to line-of-sight transmission, these telegraphs required many relay stations and personnel; moreover, they worked slowly compared with the speed of electricity; and most shut down at night. These were the performance shortcomings identified by the many proponents of *electrical* telegraphs.

Visions for an *electrical* telegraph originated in the middle of the eighteenth century (Fahie 1884; Schiffer et al. 2003). Surprisingly, a handful of inventors actually built prototypes employing electrostatic generators and Leyden jars (the latter were the first capacitors, which store an electric charge); none was replicated. Such prototypes continued to be built into the early nineteenth century, but these designs eventually were selected against in favor of telegraphs employing electromagnetism and batteries.

After Oersted's surprising discovery in 1820 that an electric current, flowing through a wire, created magnetism that could, for example, deflect a compass needle (Dibner 1961), researchers appreciated the possibility that electromagnetic apparatus could produce action at a distance, capable of carrying information. Thus, in several nations, electrical researchers conjured up visions of *electromagnetic* telegraphs; this was, after all, an invention that appeared "obvious" (Barlow 1825:105) – at least in principle.

The development of prototype electromagnetic telegraphs received added impetus after Joseph Henry's redesign of Sturgeon's electromagnet in 1831 (Henry 1831), and the invention, beginning in 1836, of various "constant batteries" by J.F. Daniell, W.R. Grove, and others. Though hardly constant in output, the new batteries needed maintenance less often than earlier designs, and so could power a telegraph for longer periods (Meyer 1972).

Prototype telegraphs included, at a minimum, technological objects that met the following use-related performance requirements: (1) a transmitter for encoding information into electrical signals, (2) a receiver, using an electromagnet, for decoding the electrical signals and displaying the resultant information visually or acoustically, (3) a battery for supplying electricity to activate the electromagnets, (4) one or more wires for connecting the transmitter and receiver, and (5) a code-book for enabling translations at both the sending and the receiving stations.

In attempting to realize these performance requirements, inventors generated many prototype telegraphs in the 1830s and 1840s whose technological objects varied greatly (the best book-length sources on these inventions are Preece and Sivewright 1891; Prescott 1888; Sabine 1869; Schellen 1850; Shaffner 1859). For example, some systems used one wire, while others used two or five, and a few many more; some employed a needle indicator on the receiver, while others employed a printer or sounder; some used codes representing letters and numbers, while others were keyed to sentences in a telegraphic dictionary. And transmitter designs were equally diverse. Some systems worked reliably, others did not, but many achieved the ability to send and receive information over many miles.

During the telegraph's early years, patents were already being treated in many nations as a form of intellectual property that could be sold, leased, or otherwise managed (Cooper 1991; Post 1976). Ambitious inventors throughout the West patented their systems, along with thousands of technological objects, which furnish a stunning record – partial, to be sure – of the invention cascades occurring during the telegraph's first decades (e.g., United States Commissioner of Patents 1883; Great Britain Patent Office 1859, 1874, 1882).

With functioning prototypes and patents, inventors can sometimes acquire modest funding and entrepreneurial expertise to continue development. And so it was with some early telegraphs. In the United States, for example, Samuel Morse teamed up with Alfred Vail whose father was a successful manufacturer (on the early history of the Morse telegraph, see Morse 1973; Taylor 1879; A. Vail 1845, J. Vail 1914). Other inventors, including Wheatstone and Cooke in England and Siemens in Germany, also obtained support, generated new technological objects, and brought their telegraph systems to market.

Technological Display

Inventors easily come to believe that their prototypes, usually assembled of jury-rigged components in a laboratory or workshop and often operating erratically, are technically feasible. Promising prototypes occasionally attract the first backers, but deep-pocket capitalists, potential manufacturers, governments, and a curious public (perhaps tempted by stock offerings) require a convincing demonstration. In the technological display process the CTS is exhibited, usually in an elaborate show-and-tell, to an outside and sometimes skeptical audience.

Because technological display must impress mostly nontechnical people, visual performance characteristics of the technological objects become critical. Indeed, the appearance of the system contributes, symbolically, to demonstrating the inventor's technical competence.

The Morse telegraph provides a dramatic example of technological display. In the telegraph's first major show-and-tell for a nontechnical audience, which took place in February 1838 in Washington DC, Vail and Morse – exploiting a connection in Congress – were able to garner an august group of onlookers that included President Martin van Buren, members of the House Commerce Committee, and heads of executive-branch departments (Vail 1845:78). These men witnessed the transmission of information through two spools of wire, each five miles long, between committee rooms in the Capitol. In preparation for this display, Vail had given the electrical parts a finished appearance. Moreover, this was the first Morse telegraph that transmitted all information – numbers and letters of the alphabet – as dots and dashes, which were recorded by a fountain pen bobbing up and down on a spring-driven, paper-covered drum. Needless to say, it was a most impressive electrical and visual performance.

Demonstrating "Practicality"

The CTSs constructed for technological display are often essentially complete systems but built on a very small scale. What is more, they are usually presented in an environment more benign than would be encountered in real-world operation. Thus, even after a successful show-and-tell, many questions remain about the system's performance characteristics. That is why a large-scale demonstration is sometimes needed to convince others that the system is "practical" (the usual nineteenth-century term was "practicable").

Practicality is taken here as the judgment that outsiders render after witnessing, or learning about, a full-scale demonstration. In capitalist-industrial societies, such judgments are based on critical performance characteristics, such as cost estimates for building and operating the CTS, the likely reliability of the system and its components, and how well it performs symbolically in specific activities and in relation to particular groups. These assessments often lead to forecasts about the size and socio-economic composition of anticipated markets. A judgment of practicality may liberate resources for replicating the system; a negative judgment may presage the CTS's demise.

The successful demonstration of practicality does not, however, ensure that the CTS will be brought to market, for replication depends on contextual factors far beyond the inventor's control, such as political enablers and inhibitors, and the availability of capital. On the other hand, inventors with considerable resources of their own may ignore negative judgments based on market forecasts and manufacture the invention anyway.

With the technical feasibility of Morse's system apparently not in doubt, Congress furnished Morse in 1843 with $30,000 to build a telegraph line connecting the Capitol, in Washington DC, with the railroad depot on Pratt St. in Baltimore, Maryland – a distance of about 40 miles (Vail 1845). As Morse and other inventors began constructing demonstration telegraphs, they encountered countless problems, which occasioned many invention cascades. For example, Morse began installing his line underground, believing that it would be more secure from vandals and sabotage than an overhead arrangement. However, after laying just 10 miles of line, Morse had already spent half the government grant; more troubling still, he found that the cable was defective. He abandoned the original plan and resorted to suspending the wires from wooden poles. In England and Germany inventors devised different – and somewhat more successful – designs for underground cables along with their diverse designs for aboveground lines.

Aboveground lines were cheaper, but they too required new inventions, such as appropriate poles (wood or metal), for suspending the wires, insulators to electrically separate the wire from the pole, rain and snow shields, methods of treating wooden poles to retard decay, treatments of the (usually iron) wire to deter corrosion, techniques for splicing wires, and new kinds of electrical connectors. For each of these performance requirements, inventors devised numerous technological objects. And, to furnish electricity for their telegraph lines, Morse and other system builders could choose among many dozens of battery designs, some invented for telegraph use.

Once a demonstration telegraph line was up and running, performance characteristics relevant for judging practicality could be assessed, including rates of transmission and operating costs. Observers judged Morse's line a rousing success. For example, the Secretary of the Treasury wrote to the Speaker of the House that "the perfect practicability of the system has been fully and satisfactorily established" (quoted in Vail 1845:98). Comparable large-scale demonstration projects in Europe of quite different telegraphs led to similar judgments.

Replication

On both sides of the Atlantic, substantial resources were poured into building telegraph systems. In European countries, some of whose governments underwrote telegraph replication, this new communication system became, like the semaphore telegraphs, a political technology (Nickles 2004). For example, the far-flung British empire was governed telegraphically from London as soon as submarine cables united the continents in the early 1870s (Headrick 1981). In the United States, however, the telegraph was proliferated by private companies, and some even competed against Morse with alternative technologies (Reid 1879). Despite differences in political and economic contexts, comparable invention cascades arose on both sides of the Atlantic during replication and in subsequent life-history processes.

In the replication process, new activities arise for manufacturing multiple instances of the technological objects. In turn, these manufacturing activities have critical performance requirements that lead to new tools, sometimes even to specialized workshops or factories. The result is usually a plethora of inventions. Moreover, as new tools are winnowed in manufacturing activities, the CTS's technological objects themselves sometimes undergo design changes to enhance ease of manufacture.

As telegraph companies were formed in the United States and in other nations, demand surged for telegraph components. Not only were new companies formed to manufacture transmitters and receivers, but established makers of wire, electrical instruments, and so on scaled up their operations (for an overview, see Israel 1989). In companies old and new, manufacturers tried out countless inventions that might promote rapid and efficient production. For example, to make wire to demanding specifications and in unheard-of quantities required new production machinery. Diverse machines were also invented for applying insulation to wires and for winding wire on electromagnets.

Marketing and Sales

To facilitate marketing activities, wholesale and retail, inventors devised lavish brochures, fancy demonstration devices, tokens, and so on. For decades, telegraph companies and manufacturers of components used these kinds of symbolically loaded objects to hawk their wares at electrical exhibitions and world fairs. Likewise, offices where people could send messages had to be furnished not only with telegraph equipment and new writing technologies (such as forms), but also with characteristic trappings, such as signs and furniture, that could help people to symbolically distinguish a telegraph office from other places of business.

Installation

As people begin to gain experience in installing the system, still more invention cascades arise. Installation-related inventions are generated to solve recurrent problems and also to routinize work, reduce labor requirements, and conserve materials.

To assist in installing aboveground lines, machines were invented that could stretch the wire to an appropriate tautness between the poles. Achieving good insulation of the wires where they attached to the poles led to dozens of insulator designs, in which inventors strived, for example, to increase electrical resistance, durability, and ease of installation.

Additional invention cascades arise when the CTS is installed in a different environment because new critical performance characteristics can come into play. Attempts to lay telegraph lines under rivers, across the British Channel, and eventually across oceans created seemingly endless invention cascades. Submarine lines required a waterproof, heavily insulated, good conducting, and strong cable that could be laid reliably. A great many people invented cables aimed at achieving acceptable values of these performance characteristics.

Accompanying the efforts to lay ocean cables, which began around 1850, were many inventions for storing the cable aboard ships and paying it out. This machinery was complex, requiring constant monitoring of the tension on the cable as well as brakes that could be applied firmly but gradually so as not to cause a break (Dibner 1959). Eventually, ships equipped with special-purpose equipment were built for cable work (Finn 1973).

Entirely new kinds of electrical instruments, such as Thomson's mirror galvanometer, enabled faint signals to be detected and allowed installers to pinpoint the location of breaks in the cable or weak places in the insulation as it was being laid.

Use/Operation

As users begin to acquire familiarity with a CTS, new use-related performance characteristics, even some unanticipated by manufacturers, may become critical. Indeed, inventions made by users are sometimes incorporated through feedback into the CTS's design (Oudshoorn and Pinch 2003). For example, people discovered quickly that lightning could wreak havoc with the telegraph, and so they invented protection devices; some lightning conductors were attached to insulators, while others were emplaced on the poles or telegraph stations.

The process of use may involve varied activities and social groups, each with different performance preferences. In the case of the telegraph, at least two user groups contributed to invention cascades: (1) telegraph operators and (2) customers (people who sent and received messages). Throughout the telegraph's first decade, operators crafted endless varieties of transmitters, receivers, batteries, and so forth in order to improve ease of use and reliability; Thomas Edison was the most famous member of this group (Israel 1998). Consumers, actual and potential, can contribute to invention cascades by calling attention to new applications (see "Functional Differentiation").

Perhaps, the most important source of invention cascades during use is growth of the system. As a CTS is forced to accommodate more users or a greater intensity of use, scalar effects can degrade core performance characteristics. Solving these problems necessitates expansion of the system, either by building more systems identical to the original or by changing the CTS's technological objects to increase its capacity. Both solutions were adopted as demand for telegraph service rose sharply during the middle of the century. In addition to building more lines, or adding new wires to old lines, inventors such as Edison came up with countless technologies for sending two or even four messages on the same wire.

Maintenance

As installed systems begin their uselives, varied maintenance activities are necessitated. Some are easily predicted or become apparent quickly because they occur often; others may not be evident until the system has been in use for some time. Both high- and low-frequency maintenance requirements can occasion invention cascades.

Refurbishing telegraph batteries was a predictable and high-frequency maintenance activity, one that was distasteful to telegraph operators because batteries contained acid. Replacing electrodes and renewing the acid was a messy and dangerous job. Not surprisingly, efforts to invent more easily maintained batteries created a constant flow of inventions, some offered by telegraphers themselves.

Infrequent maintenance activities, such as repairing damage to poles and lines after an ice storm, also stimulated invention cascades. In particular, the need to locate breaks in the line and to troubleshoot malfunctioning equipment led to new instruments and standard units for measuring voltage, current, and resistance.

The repair of submarine cables, damaged by animals, anchors, contact with rocks, and other causes, gave rise to rich invention cascades. To recover the ends of a severed cable, for example, required new kinds of grappling hooks. Once the cable was captured, of course, the free ends had to be joined by special splicing technologies – the source of another invention spurt.

Functional Differentiation

After replication, a CTS often enters a visible public realm where people in diverse communities of practice consider using it for their own activities. The process of adapting the technology for new activities sets off more invention cascades (on the process of technological differentiation, see Schiffer 2002). The new systems that result could, for purposes of analysis, be treated as entirely new CTSs and studied in their own right.

In the case of the telegraph, many new invention-stimulating functions materialized early on. Among the first were railroad telegraphs for signaling the locations and conditions of trains to the dispatcher (Langdon 1877). Inventors came up with varied transmitters for use on trains and others that could tap into the line anywhere along the tracks. Eventually, there were alternative designs for trackside, electrically controlled signaling systems that responded to the movement of trains and to orders from dispatchers.

Another new application was the municipal "fire-alarm telegraph," developed simultaneously, and probably independently, in the United States and Germany around 1850 (Anonymous 1862; Channing 1855). A fire-alarm telegraph furnished fire stations with timely information on the location of fires. Throughout cities, fire-alarm boxes containing telegraph transmitters were placed along streets. When a signal announcing the outbreak of fire in a particular district arrived at the central station, a dispatcher would alert the closest fire brigade, also by telegraph. These systems stimulated a flurry of inventions that, among other performance characteristics, (1) enabled anyone to set off a fire alarm, (2) provided the dispatcher with a display indicating which alarm had been activated, and (3) permitted fire brigades to receive alerts.

Visions of other specialized telegraph systems also provoked invention cascades, including hotel "annunciators," through which guests could signal their needs to staff; burglar and fire alarms in homes and businesses; stock tickers for connecting offices and homes to stock exchanges; and portable military telegraphs that could be moved along with troops.

Operationalizing the Cascade Model on Archaeological Cases

This section suggests that the cascade model can become a useful archaeological tool for investigating CTS-related invention processes in diverse societies.

Applicability of the CTS Construct

Inquiring minds doubtless wonder whether CTSs are even present in the societies that most prehistorians study. Employing the flexible definition of CTS presented above, many technologies in small-scale societies appear to conform. For example, the bow and arrow is a CTS, composed of several separately functioning technological objects that help to achieve the system's core use-related performance characteristic: the ability to aim an arrow and launch it at a sufficient velocity to wound or kill an animal (see Hughes 1998). Domestic cooking technology might be a near-universal CTS, consisting of technological objects, such as containers, utensils, ingredients, and a heat source, which functions to transform edible substances into culturally appropriate meals. Some ritual technologies, recreational technologies, enculturative technologies, political technologies, soil- and water-control technologies, plant-cultivation and animal-husbandry technologies, and the like could also be regarded as CTSs. In view of the construct's definitional flexibility, I submit that CTSs should be identifiable in virtually all societies.

The next issue is whether the cascade model's life-history processes are applicable to CTSs in small-scale societies. It would appear that the basic set of processes – that is, creation of a prototype, replication (or manufacture), use, and maintenance – is general enough to be nearly universal. As in the telegraph case, the archaeologist can add other processes to the basic set.

Another issue is whether the development of CTSs in small-scale societies gives rise to invention cascades. In principle, performance problems should emerge during life-history processes in the development of any CTS – regardless of societal context. Consider once more, in a thought experiment, the bow and arrow. Inventors could acquire the vision for this CTS from many sources: thinking about new ways to hunt, watching hunters in another society, or even handling a bow and arrow made elsewhere. Regardless of the vision's origin, attempts to realize it might have stimulated trials with new materials that had to be worked and assembled in new ways. Moreover, the creation of bow-and-arrow prototypes likely entailed the invention of new tools and processing techniques. And the bow and arrow's use on different game animals might have disclosed additional performance problems. It is doubtful that ancient hunters would have arrived at completely workable designs on the first try. Probably there were flurries of inventions, which yielded along the way unsuccessful technological objects. Moreover, if bows and arrows acquired important symbolic functions, then relevant visual performance requirements would have stimulated still more invention cascades. If this thought experiment is indicative, then one would expect that creating even the simplest CTSs in prehistory resulted in some invention cascades. The alternative position, it would appear, is that prehistoric inventors were omniscient, able to predict unerringly which technological objects would allow a CTS to carry out its utilitarian and symbolic functions.

Seemingly, the cascade model is sufficiently general and flexible to be operationalized on the archaeological record of small-scale societies. Yet, there remains a pressing question: in applying and evaluating the model, how might the archaeologist proceed? The answer consists of a thumbnail sketch of possible research activities. The list that follows is not a recipe, however, for it is likely that provisional findings will give investigators a basis for repeating the research activities in varied sequences.

One begins by identifying a CTS. Let us take, for purposes of discussion, "canal irrigation among the Hohokam," an archaeological culture that occupied a large part of southern Arizona between about AD 500 and 1450 (inspiration for this CTS comes from Ackerly et al. 1987; Dart 1989; Gumerman 1991; Haury 1976; Huckleberry 1999).

The investigator next defines the CTS in behavioral terms by specifying a small set of core performance requirements that would have permitted a prototype system to function. Thus, a riverine canal irrigation system has to convey water from a river to cultigens and enable farmers to control the amount of water reaching individual fields.

Using life-history processes as analytic units, the archaeologist specifies the kinds of performance problems that would have emerged during development. In attempting to solve these problems, farmers qua inventors would have generated invention spurts to yield technological objects having suitable performance characteristics. Replication, for example, probably required durable digging implements, major and minor canals capable of handling the usual flows, devices for easily and reliably controlling the flows to each field, and fields whose design promoted ease of irrigation. Farmers also might have come up with inventions that enabled the laying out of suitable canal routes. To handle maintenance problems, farmers likely would have devised artifacts that could remove accumulated sediments, patch weak or eroded places in canals, repair or replace control devices, and rehabilitate washed-out fields. In extreme cases, such as the aftermath of a huge flood, large parts of the system might have been rebuilt with new canals that had differing lengths, grades, and cross sections. To deal with salinization of fields, farmers could have tried out new crops to find salt-resistant varieties. Expansion of the system might have necessitated additional inventions, such as new kinds of canals as well as technologies for lengthening and raising the capacity of old ones. If the canal system acquired new functions, such as furnishing water for domestic consumption and clay for making pottery, new performance requirements could have stimulated further invention cascades.

In inferring the performance problems that emerged in a developing CTS, the archaeologist must understand in detail how the system would have worked. To acquire such knowledge – that is, "techno-science" (Schiffer and Skibo 1987) – one can exploit modern engineering literature and expertise, conduct experiments, and draw upon ethnographic, ethnoarchaeological, and historical information. This high level of understanding (*not* displayed in the canal irrigation example) lays a foundation for inferring – from archaeological evidence – the technological objects that seemingly had the requisite performance characteristics for taking part in specific life-history activities.

After inferring which artifacts were likely to have been part of the CTS, the archaeologist partitions them into sets according to life-history processes. Next, the time–space distributions of the members of each set are delineated as precisely as possible. The archaeologist can then scrutinize these distributions for any patterning that might be interpretable as invention cascades, paying special attention to variants that apparently were unsuccessful. For example, suggestions of invention spurts may come from diversity in canals, especially those that went nowhere, were damaged without repair, or were abandoned and replaced almost immediately after construction. Repair episodes and other modifications that appear to have been used for only the briefest period might also help to pinpoint invention spurts. In attending to unsuccessful variants as products of invention cascades, the archaeologist might be able to make sense of variability that was previously obscure or ignored.

Although it would be desirable to make predictions about the temporal patterning of technological objects (aggregated by life-history processes) in the development of CTSs, such an effort would be premature in light of current knowledge. After all, it can be expected that different CTSs will have different developmental trajectories. Moreover, predictions are rendered difficult for CTSs that underwent relatively continuous invention cascades in response to changing contextual factors (e.g., automobiles, the electromagnetic telegraph, and perhaps canal irrigation systems). After all, new technological objects could be invented early, late, or throughout the CTS's life history, precluding *general* predictions about the order of specific invention spurts. Clearly, the development of each CTS must be examined empirically. In the future, however, archaeologists might be able to formulate some generalizations after conducting comparative studies of invention cascades in diverse CTSs. Such studies might also lay a foundation for subdividing the general behavioral context – CTS – into varieties that are characterized by distinctive developmental trajectories and thus temporally patterned invention cascades.

Discussion

As noted elsewhere (Schiffer 2002), by employing behavioral models the investigator can establish a foundation for constructing historical narratives of technological change. Thus, after doing an analysis guided by the cascade model, the archaeologist could fashion a reader-friendly narrative about the CTS's development. The structure and content of that narrative, however, would be underdetermined by the cascade model. This leaves ample room for archaeologists who prefer, for example, agency, social construction, or evolutionary explanations to craft their own narratives on the behavioral foundation. Indeed, because behavioral models direct attention mainly to proximate contexts, one can create narratives that invoke more distant, but still causally relevant, contextual factors.

It should be apparent that the cascade model's demanding inferential requirements could preclude its literal application in many cases. For example, the technological objects of canal irrigation systems, especially the canals themselves, are difficult to date (but see Eighmy and Howard 1991). Nonetheless, even in such difficult cases the cascade model can serve a useful purpose by calling attention to hitherto neglected and unexplained kinds of archaeological variability, such as the unique variants – from canals to firepits to decorated sherds – that do not conform to established types. These variants are often treated as inexplicable idiosyncratic variation, dropped into "other" categories and promptly forgotten. *Some* of these artifacts and features could have been failed variants generated by invention cascades. Merely asking questions about the sources of such variability might provide an inductive entrée into the invention cascades of a CTS.

Implications of the Cascade Model

The cascade model provides potentially fruitful ways to conceptualize some processes of technological change.

CTSs and Material Technologies

A CTS can, and often does, include technological objects made by artisans working in different material technologies. As examples, the telegraph incorporated objects of metal, wood, and glass, and the CTS of domestic meal preparation can include ceramic, chipped stone, and wooden objects – not to mention plants and animals. Thus, a CTS's invention cascades can lead to new variants in different material technologies. Once a CTS has been delineated, the investigator attempts to pinpoint the performance problems that provoked invention cascades in diverse materials.

By the same token, temporal change in the objects of a particular material technology might have resulted from invention cascades in different CTSs (cf. Sillar and Tite 2000:14). For example, let us consider the continuous changes in Anasazi ceramics of the American Southwest that took place from about AD 600 to around 1400 (e.g., Cordell 1997; Chap. 3). Such changes doubtless resulted from altered performance requirements in several CTSs, such as ritual technology, domestic meal-preparation technology, and feasting technology. Potters responded by inventing vessels having an amazing variety of pastes, forms and sizes, and surface treatments, some of which were replicated in large numbers. It might be productive to consider the possibility that practitioners of a given material technology were inventing objects that were supposed to interact in different CTSs.

If a CTS can foment invention cascades in several material technologies and if a material technology can create new variants for several CTSs, then we need to rethink analytical strategies that treat material technologies as autonomous behavioral phenomena. This discussion also implies that any given material technology could have been invented, in various places, in response to the development of different CTSs (Rice 1999 has made this argument for pottery origins).

Necessity as the Mother of Invention

The cascade model also invites reconsideration of the old question: Is necessity the mother of invention? (For a discussion of this question from the standpoint of evolutionary ecology, see Fitzhugh 2001.) Setting aside the issue of whether the telegraph was a response to needs, once efforts were underway to develop a functioning system, inventors had to devise new technological objects necessitated by the system's core performance requirements. Although these requirements could be met in many ways, all functioning telegraph systems employed some constellation of new objects. Likewise, developing a functioning Hohokam canal system required the invention of new technological objects, including water-control devices, canals of several kinds, and irrigable fields.

We suggest that any CTS has core or critical performance requirements, emergent during life-history processes, that determine its functional "needs" (utilitarian and symbolic). Meeting these needs, through invention cascades, entails the creation of new technological objects. Thus, *in functional terms*, the inventions spawned by a given CTS result from necessity: if the CTS is to operate as a system, then these inventions must be made. Given the apparent prevalence of CTSs, one could argue that necessity is the mother of a great many inventions (for a contrary view, see Basalla 1988).

Developmental Distance

Although the vision of a new CTS is sometimes obvious to knowledgeable members of a society or community of practice, far from obvious are the forms, specific functions, performance characteristics, and manufacturing processes of the new technological objects needed for the system's replication, operation, and maintenance. Indeed, the vastness of the development enterprise often becomes apparent only as inventors encounter the innumerable performance problems that emerge during life-history processes.

This idea is shown in the writings of countless visionaries, from Leonardo da Vinci onward, which indicate that machine-powered human flight was an idea that cropped up often. In the nineteenth century, especially after the advent of the railroad and steamship, the vision of self-propelled road vehicles also occurred to many people throughout the Western world. Both visions stimulated invention cascades that resulted in prototypes, but only the automobile achieved acceptable values of core performance characteristics before 1900. Neither CTS was widely replicated and adopted until after many invention cascades led to new technological objects that solved myriad "little" performance problems, such as the ability to control an inherently unstable aircraft or to cool an internal combustion engine.

In order for a CTS to move from a vision – obvious or not – to a replicated technology, its inventors must traverse a certain "developmental distance." That is, they must generate cascades sufficient to produce variants that can help solve the entirety of emergent performance problems. Some developmental distances are short, perhaps because a functional CTS can be cobbled together from technological objects already invented and replicated in other contexts. Sometimes only a few performance problems arise, and so generate only a few spurts of invention. In other cases, developmental distances are lengthy, such as those attending the emergence of telegraph, automobile, and riverine canal systems. A large developmental distance usually compels an enormous investment of human and material resources in inventive activities. As already noted, the societal context looms large in determining whether and in what manner the necessary resources can be devoted to the project.

In small-scale societies, there might have been a lack of sufficient resources for bridging huge developmental distances quickly or at all. At the very least, scheduling conflicts can preclude the diversion of human labor into inventive activities which, as Fitzhugh (2001) reminds us, usually have an uncertain outcome. Consider the case of domestic structures used for storage and habitation, a CTS among the Anasazi of the American Southwest (Cordell 1997). During a period lasting many decades, the Anasazi transformed their structures from pit houses and sundry storage facilities to mostly aboveground, masonry pueblos encompassing both habitation and storage functions. Evidently, changing contextual factors in Anasazi society, such as community reorganization, lengthier stays by households in one settlement, and longer settlement occupations (perhaps set in motion by larger village populations and increasing dependence on agriculture), gradually established new core performance requirements for dwellings and storage facilities (Cordell 1997; Gilman 1987; McGuire and Schiffer 1983; Whalen 1981). Regardless of the causes, inventing the new technological objects (and their manufacturing processes) appears to have entailed a considerable developmental distance.

Remarkably, the invention cascades that contributed to the development of pueblo structures left obtrusive traces in the archaeological record. "Transitional" Anasazi structures were characterized by diverse building techniques and designs, which testify to invention cascades that we know – thanks to tree-ring dating – played out over many decades. This lengthy period of experimentation, which relied on efforts spread over a large region, furnished the Anasazi with reliable information on the performance characteristics of various structure designs, from which they eventually selected the pueblo, which combined both storage and dwelling. One could argue that, had the selective pressures exerted by contextual changes been more insistent, the Anasazi might have been unable to marshal resources needed to traverse the developmental distance *quickly*.

Indeed, one can imagine that the failure to span a large developmental distance rapidly – for example, creating a new agricultural technology in the face of pressing demand for more food or a rapidly deteriorating environment – might have led to other behavioral changes, such as emigration, new kinds of regional organization, modified exchange networks, or violence. It would be unwise to assume that all societies had the resources to reach across large developmental distances in a timely manner. Perhaps many of the gradual technological changes so prevalent in prehistory merely reflect those occasions when there was a good match between the severity of selective pressures and the capacity of traditional societies to generate invention cascades and thereby respond with a functioning CTS.

In one surprising respect, Anasazi structures and electromagnetic telegraphs seem remarkably similar. As CTSs involving great developmental distances, both were built by the pooling of numerous small inventions, generated by cascades, that had been made over several decades by many inventors working in many places. Perhaps this pattern is common.

Conclusion

Drawing upon the richly documented history of the electromagnetic telegraph, a model of invention cascades was presented that applies to complex technological systems (CTSs). The model's key premise is that performance problems emerging during a CTS's development stimulate sequential invention spurts – cascades – that can be conveniently studied in relation to life-history processes. The minimal set of life-history processes, which should apply to most CTSs, is making a prototype, replication or manufacture, use, and maintenance. Depending on the CTS under investigation, the archaeologist may subdivide these processes and proliferate others. In principle, this model should be applicable even to the smallest-scale human societies studied by archaeologists. The cascade model, however, is just one of many models that we require for understanding the variety of invention processes prevalent in human societies.

It should be emphasized that the building of general models does not conflict with the creation of deeply contextualized historical narratives. Beginning with their earliest writings, behavioralists have acknowledged the importance in archaeology of both generalizing and historical research strategies (e.g., Reid et al. 1975), and have also crafted lengthy narratives of technological change (e.g., Schiffer 1991; Schiffer et al. 1994a, Schiffer et al. 2003). However, archaeologists have seldom exercised the generalizing research option when studying invention. This leaves the door open for devising new models and theories that can complement narratives by implicating widespread invention processes operative in specific behavioral contexts, such as CTSs. By constructing and evaluating general models of invention processes, archaeologists can make significant contributions to the study of technological change.

Chapter 6
Ritual Performance: Ball Courts and Religious Interaction[1]

Religion and ritual are often considered part of the ideological system that is outside the realm of technology and thus beyond the scope of archaeological consideration. The core belief of our model is that any behavior, ritualistic or otherwise, will involve material culture and thus can be explored with the performance-based life history approach. Walker (1995a, 1995b, 1996, 1998, 2001) has been successful in demonstrating that ritual behavior, like any human behavior, shapes the life histories of artifacts. One of his most significant contributions has been to demonstrate a stratigraphic approach to ritual in which quite ordinary evidence such as trash fill, roofing materials, animal bones – items and deposits usually not considered ritual in nature – are used to inform on ritual behavior.

In this chapter, we combine this stratigraphic, deposit-oriented model with the performance-based approach to inform on the ritual activity during the Animas Phase. We argue that ritual activity – in particular, ball court ritual performance – explains in large part the social interaction between the site of Joyce Well, as well as other Animas Phase sites in the Southwest, and the important site of Casas Grandes of Chihuahua, Mexico (see Walker and Skibo 2002).

Joyce Well is an Animas Phase (AD 1200–1400) pueblo located in the extreme southwest corner of New Mexico (Fig. 6.1). The pueblo contains an estimated 200 rooms (Fig. 6.2) and was excavated on two occasions, first by School of American Research in 1963 and then in 1999–2001 by the La Frontera Archaeological Research Program (Skibo et al. 2002). The Animas Phase pueblos, like Joyce Well, represent a northern frontier of the Chihuahuan culture horizon centered at Casas Grandes (also known as Paquimé). Communities in this regional system possessed similar architecture, artifacts, and rock art, which suggest a shared series of beliefs and practices (e.g., Di Peso 1974; Ravesloot 1979; Schaafsma 1997). Connections to Paquimé seem unambiguous, but the nature of the relationship – economic, ritual, or political – between these outlying communities and the powerful city located 120 km to the south remains hotly contested (Schaafsma and Riley 1999; Whalen and Minnis 2001).

[1] This chapter is cowritten by William H. Walker.

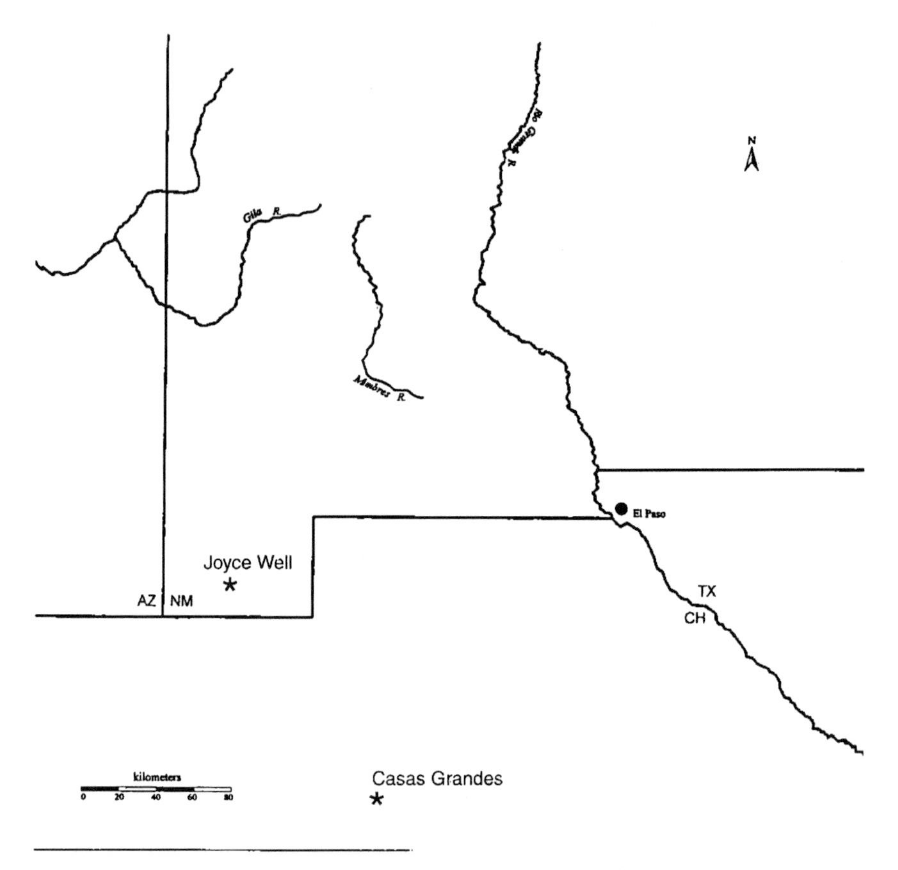

Fig. 6.1 Location of the Joyce Well site in southwestern New Mexico

This chapter reports on the results of the excavation and analysis of the Joyce Well ball court, which is located 80 m north of the pueblo (Fig. 6.3). The role of ball courts in regional interaction is also discussed, and we provide some preliminary information on two other ball courts located within 11 km of Joyce Well (Culberson and Timberlake). Moreover, activities that took place within the ball courts at the Joyce Well, Culberson, and Timberlake communities are explored through a performance-based analysis. By focusing on the attributes of the courts themselves, the associated performance characteristics, and the ethnohistoric documentation of ball courts, we conclude that the courts had three primary functions. First, based on the accessibility, capacity, and location of the features, we argue that the courts performed as a community integrative mechanism. Second, the true north orientation of the features and their similarity in size and shape to other ball courts located to the south suggest that a celestial-based rubber-ball game was performed in the

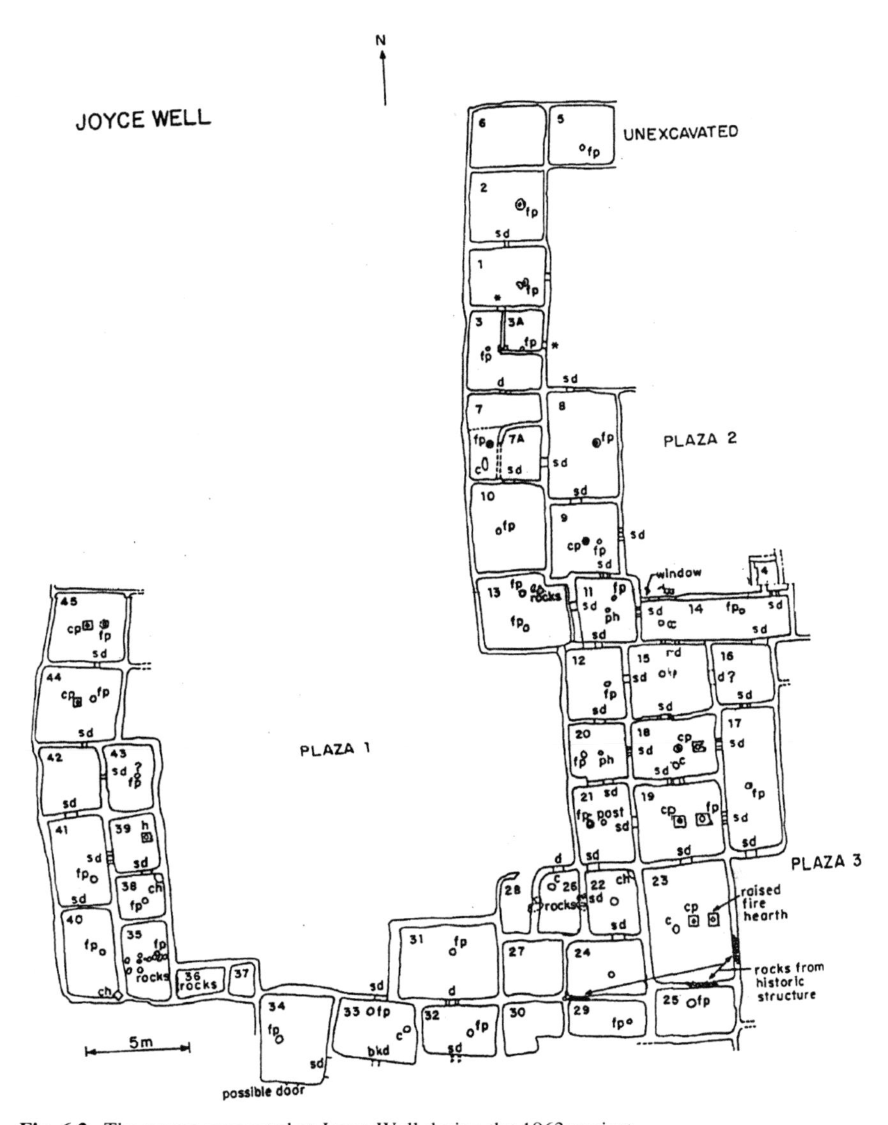

Fig. 6.2 The rooms excavated at Joyce Well during the 1963 project

courts that was related to fertility rituals. Third, and finally, we argue that the three ball court communities in the boot heel of New Mexico were part of the Casas Grandes religious interaction sphere. This is based not only on the similarity in courts between the Animas Phase sites and the courts located nearer to Casas Grandes, but also on other Joyce Well artifactual and architectural data.

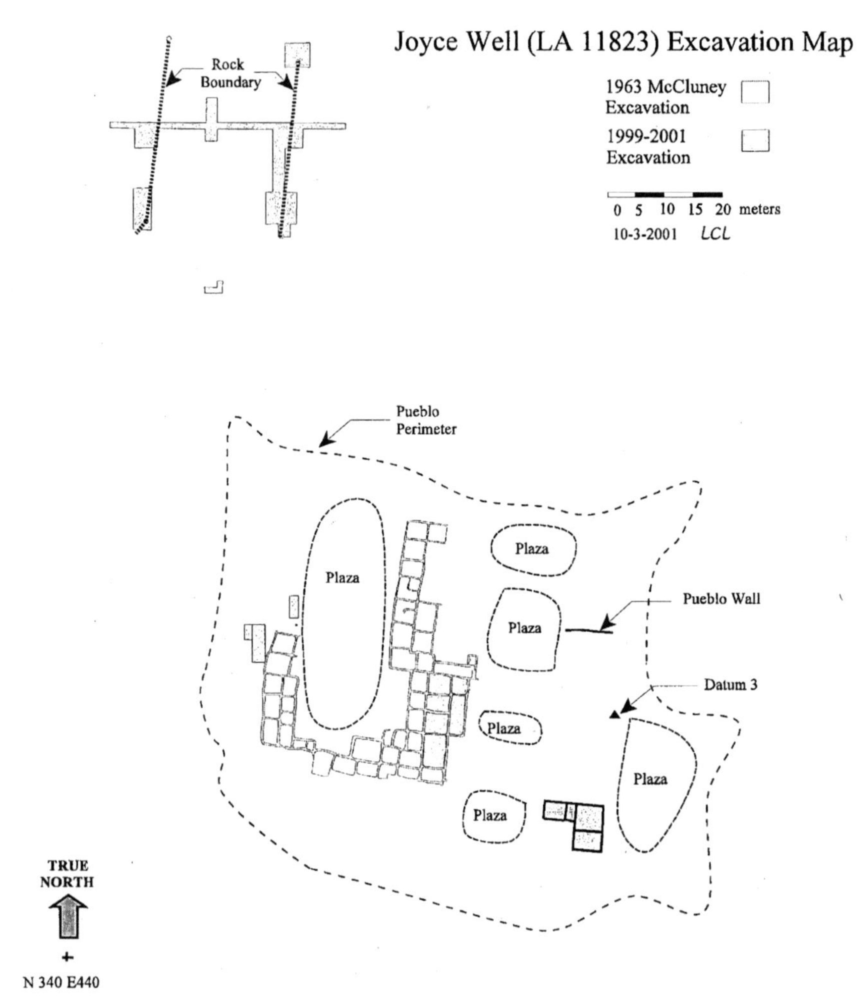

Fig. 6.3 The relative location of the pueblo and the ball court at Joyce Well

Joyce Well Ball Court

The Joyce Well ball court was not identified by McCluney's team in the 1960s. In fact, no Chihuahuan-style ball courts, which typically consist of two parallel rows of rocks and often a slight berm and interior depression, had been recognized at that time as features within the Casas Grandes regional system. Primarily through the work of Whalen and Minnis (1996, 1999; see also Fish and Fish 1999; Leyenaar 1992; Schaafsma and Riley 1999), however, many of these features have now been reported and a total of 22 have now been recorded in Chihuahua and southwest New Mexico.

The three courts in our study area are the only Chihuahuan-style ball courts located outside of Mexico. There has been some suggestion (Fish and Fish 1999:37–38) that a feature at the Ringo site (Johnson and Thompson 1963) may also be a ball court but it does not seem to be the type of court associated with the Casas Grandes interaction sphere. The feature is 26 m in diameter and generally oriented north–south, and it appears more similar to what Whalen and Minnis (1996:736) refer to as "quadrilateral structures that *could* have been a playing field" (emphasis in original). The feature is located between two room blocks, and a low adobe wall was found in the associated berm. Although there is some variability in known ball courts, the courts located in the boot heel of New Mexico (Joyce Well, Timberlake, and Culberson) are located away from the pueblo (either directly north or south), and they are bordered by either a single- or double-rock wall. We cannot rule out the possibility that a ball game was played at the Ringo site but it does appear to be out of the morphological and stylistic range of the Chihuahuan style ball courts associated with the Casas Grandes interaction sphere. Whalen and Minnis (1996) excluded these features, sometimes referred to as "corrals," from their discussion of ball courts in the Casas Grandes interaction sphere and we will as well.

The Joyce Well court was not recognized until recently because the feature is not only quite subtle and located away from the pueblo, but it is covered by a heavy tangle of mesquite, cholla, prickly pear, and other desert scrub. Prior to excavation we visited the site on a number of occasions and typically wandered about for some time within, literally, a few meters of the rock walls made invisible by the heavy vegetation. What is more, a feature of this size is difficult to comprehend when walls cannot be easily followed and only a small segment can be seen at one time.

Morphology

In order to understand the court's surface morphology, all vegetation was removed from the feature, which took approximately 20 people almost 2 days working with chain saws and hand tools. The brushing alone exposed approximately 50% more wall-rocks and it revealed more clearly the slightly depressed court center and the associated berm. The rocks used for construction, ranging in size from 10 to 50 cm, were water worn and were likely taken from the Deer Creek bed located just 100 m southwest of the feature. Ground stone fragments were also observed in the stonewalls.

The Joyce Well ball court basically consists of a slight depression bordered by two parallel rows of rocks and associated berms. A plan view of the walls was prepared, and each rock drawn to scale (Fig. 6.4). The interior dimensions (within the two walls) are 23.5 by 35 m (822.5 m^2) and, like the pueblo, is oriented approximately true north (the ball court is oriented 6° east of north). A total of 259 rocks were exposed during brushing (this total does not include rocks identified during excavation). The two parallel rows of rocks with a slightly longer west

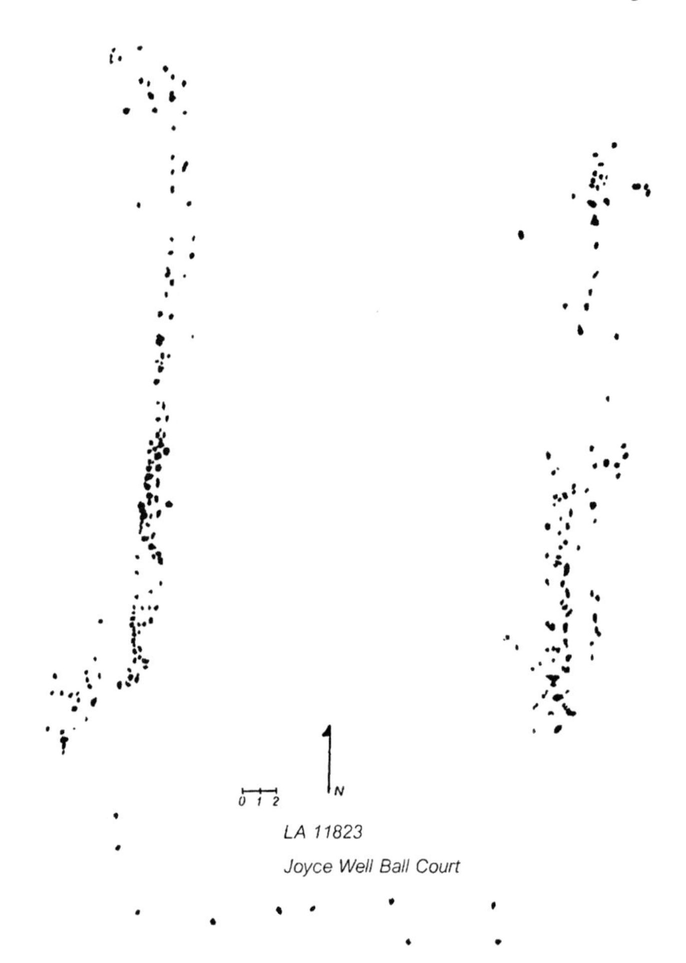

Fig. 6.4 Plan view of the Joyce Well ball court

wall are clearly illustrated in Fig. 6.4. We attribute this slight difference in length to historic disturbance, particularly on the north ends of the walls. Not only have cattle grazed on the property since the early twentieth century but the Joyce Well homestead was built on top of the pueblo and a historic road passes within 15 m of the court. The overall impression of the walls is that most of the rocks have been displaced from their original location most likely because of cattle or pedestrian traffic.

One of the least disturbed areas occurs on the west wall just south of the midpoint where vegetation served as a sand trap to protect the walls from disturbance. Here only the very tops of the rocks were exposed and they are set on edge in a double row. This pattern was repeated on various other less-disturbed segments leading us to believe that all of the walls had at one time the double-rock row pattern. The

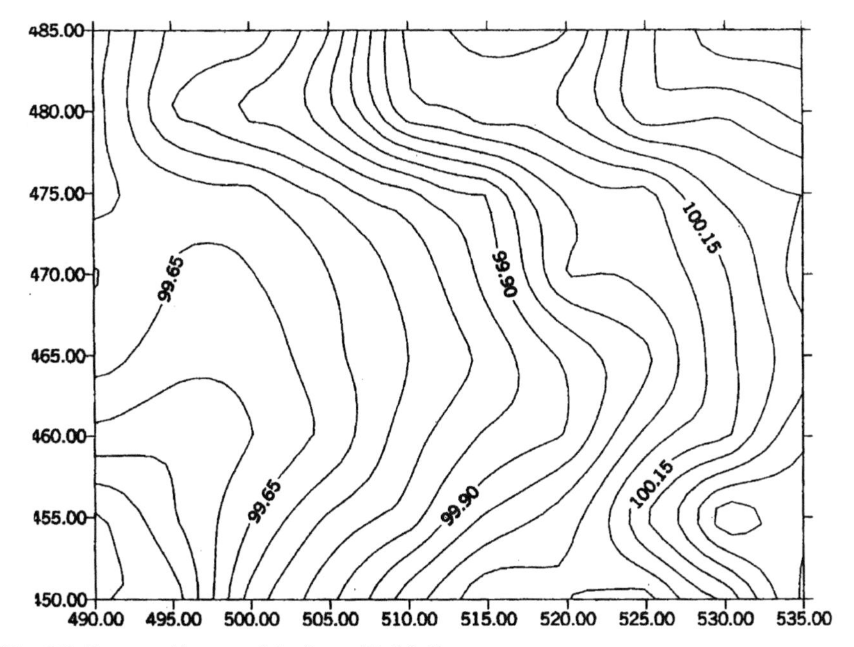

Fig. 6.5 Topographic map of the Joyce Well ball court

southern end of both the east and west walls also seem to have undergone the least disturbance owing, in part, to the larger rocks used, which are less affected by cattle traffic. At the southeast corner the walls angle outward in what appears to be a crude I shape. Here, it appears that the integrity of the original rock placement has been maintained.

A detailed topographic map of the ball court was also prepared by recording elevations at meter intervals. Fig. 6.5 illustrates clearly the central depression, east and west wall berms, and the slightly lower north and south berms. At the east–west midline of the court the difference in elevation between the berm apex and the lowest elevation of the central depression is only about 0.30 m. Nonetheless, when standing in or near the feature, the berm and central depression are unmistakable.

Excavation

The ball court excavation focused on defining the original surface of the court, exploring the architectural details of the walls, collecting associated artifacts, and searching for center- or end-court features. In all cases, the excavations were quite shallow as sterile soil was encountered between 5 and 20 cm.

An east–west trench was excavated through the court midline and the original court surface was exposed. The surface consisted of only hardpack earth; no plastering or formal floor preparations were discovered. The trench was widened at the center of the court in the search for features but none were encountered. One important question that was not resolved to our complete satisfaction was how the ball court surface and rock wall/berm intersected because the berm had such heavy disturbance from roots and rodents. At one of the least disturbed areas, however, it is clear that the floor rose gradually to the berm and there was not a formal wall that may have, for example, met the floor at 90° angle.

Segments of the east and west walls were also exposed to determine how walls were built and to investigate whether the rocks served as the only sideline marker or whether they were just what remained of a more formal wall made of adobe or wood. We found no definite evidence of elaboration of the walls beyond the double row of rocks. In some of the units, however, we did find evidence of a thin adobe-like lens that might be the melted remnants of a low adobe bench. Our experience in relocating the walls of the 1963 pueblo excavation suggests that walls exposed to the elements quickly melt away. Our best guess at this time is that the walls simply consisted of two double rows of rocks but we cannot rule out the possibility that a low adobe bench did exist between the rocks; a low adobe wall of this type would have melted away rapidly if exposed on the berm. The only reason that the adobe walls were preserved in the ruin is because many were burnt and filled before completely deteriorating.

During the excavation of the walls we did, however, find a row of buried rocks that were not associated with the previously exposed ball court rock alignment. These rocks consistently appeared about 1.5 m inside the double-row rock wall. We concluded that the rocks represent an earlier court that did not have a central depression. The remodeling of the original court involved the excavation of the central depression and the associated piling of earth that created the berm and covered the rocks of the earlier court.

Artifacts

Artifacts were not common in the feature but the frequency of ceramics and chipped stone was greatest in the berm. We have no evidence that the artifacts were the result of activities on the berm; they likely became concentrated in and around the walls as a result of the original excavation of the depression and, possibly, the routine cleaning of the court surface. A total of 464 sherds and 818 pieces of chipped stone were recovered from the ball court excavation. Lithic tools (limited to utilized flakes and scrapers) were dominated by utilized flakes, which accounted for 98% (249) of the total. An analysis of the debitage reveals a high percentage of broken flakes, which is an indicator of tool production (Sullivan and Rozen 1985). The lithic raw material is dominated by rhyolite (52.7%), which is a stone that can be picked up right on the site, followed by chert (27.2%) and obsidian (19.2%).

The obsidian is all from the Antelope Wells source found several miles west of Joyce Well. At the Antelope Wells source, obsidian occurs in the form of small nodules that can be readily collected on the surface.

A total of 464 sherds were recovered. The assemblage is dominated by plain ware body sherds as only 85 pieces were painted and 12 rims were found. Very little can be said about the ceramic material because the majority of the sherds were extremely small (mean maximum length is 23.06 mm) and abraded. In fact, 47% (216) of the sherds had extreme abrasion to the point where the surfaces were completely removed. The size distribution of the sherds also suggests that the assemblage has undergone significant trampling. Nielsen (1991) demonstrated that the size distribution of sherds after trampling with be skewed toward sizes 40 mm or less. The vast majority of the sherds are 40 mm in maximum diameter or less, and over 50% of the sherds are 20 cm or less. Combined with the fact that almost 50% of the sherds have extreme surface abrasion, it appears that our collection of sherds has undergone extreme attrition as a result of trampling. Although cattle traffic could be responsible for some of the breakage and attrition, the majority of the sherds were below the surface and would have been protected from more recent historic activity. We argue that the majority of the sherd attrition and breakage occurred while the sherds were within the ball court. These small sherds were then swept up and deposited on the berm possibly in preparation for an event.

There is no evidence that the artifacts recovered from the excavation were produced through activities associated with the games played at the ball court. We suspect that the sherds and lithic material were deposited as a result of activity in and around the ball court when it was not being formally used. It is likely that the artifact density within the ball court is no greater than that any area within 100 m of the pueblo. The concentration of artifacts within the berm, however, is likely the result of ball court-related activity. The berm was created by the removal of dirt from the center of the court and depositing it on the perimeter. Small artifacts of the type in our collection would simply have been deposited along with this dirt.

Culberson and Timberlake Ball Courts

Two other ball courts occur within 11 km of Joyce Well at the sites referred to as Culberson (LA 31050) and Timberlake (LA 54038) (see Fig. 6.6). Both of these courts appear at large Animas Phase sites as big or bigger than Joyce Well. The Timberlake ball court (Fig. 6.7) occurs on Walnut creek about 11 km northeast of Joyce Well. This court is especially interesting because, unlike the Joyce Well ball court, it is in a remarkable state of preservation, which is ironic given that the Timberlake pueblo has been very heavily potted. The court has roughly the same dimensions as Joyce Well and is oriented north (3° west of true north), but beyond that there are noticeable differences in design between the two courts. Although the court consists of two parallel rock walls, like Joyce Well, the Timberlake walls consist of a single row of rocks. In most cases, the rocks are still standing up on edge

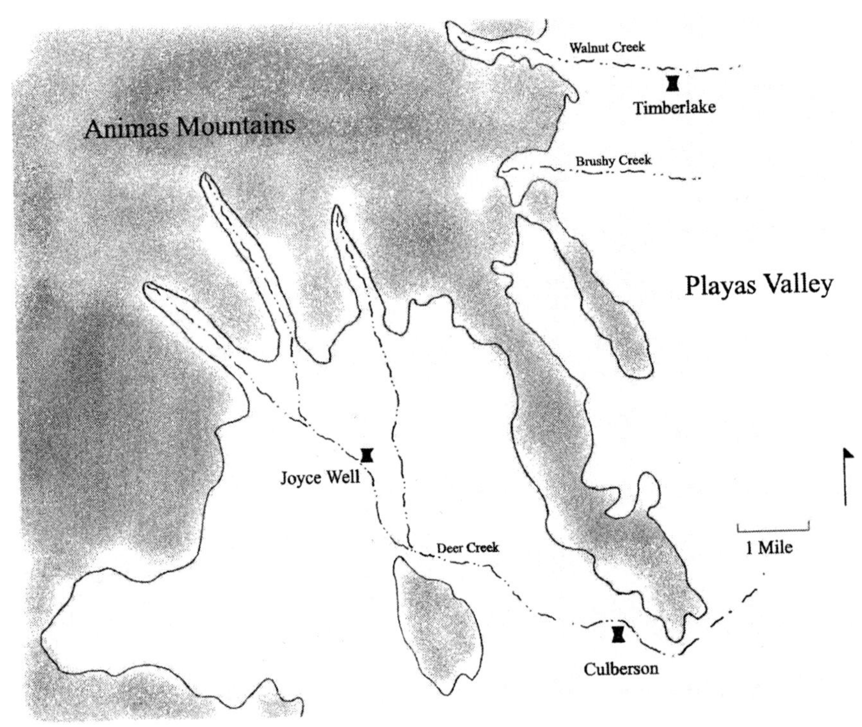

Fig. 6.6 Locations of the Joyce Well, Culberson, and Timberlake ball courts

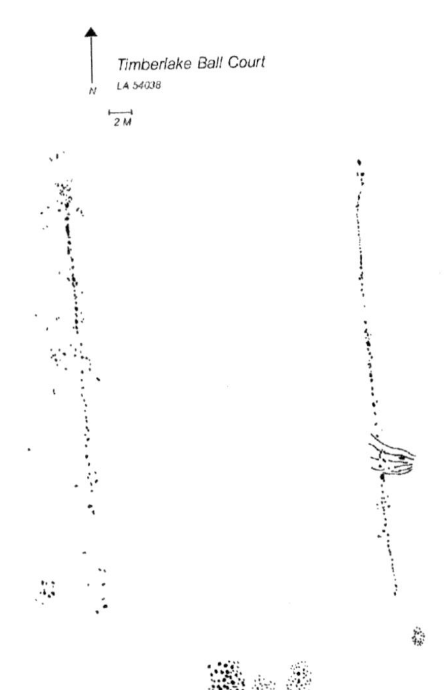

Fig. 6.7 Plan view of the Timberlake ball court

Fig. 6.8 Plan view of the Culberson ball court

as originally placed (Fig. 6.8). This court also does not have a central depression, associated berms, or I-shaped end features. Instead, there are circular rock features located at the end of the court.

The Culberson ball court (Fig. 6.8) is also located on Deer Creek roughly 7 km southeast of Joyce Well. This feature, however, has been heavily disturbed, and it is impossible to determine from the surface whether the walls were made of one or two rows of rocks. This is the smallest of the ball courts (22 by 22.5 m) but like the Joyce Well and Timberlake courts it is oriented north (4° east of true north). Like Timberlake, the Culberson ball court has no associated berms and depression, and the court is located immediately south of the ruin. The ball court at Joyce Well is located immediately north of the ruin. In all three cases, however, the pattern is identical in that the pueblo is located between the court and the stream on a N–S line.

The Boot Heel Courts

All three of the courts located in our study area are of the "simple open variety" as discussed by Whalen and Minnis (1996). They are oriented north–south and built simply by clearing a playing field and lining the borders with one or two rows of rocks. In the case of Joyce Well, a central depression was excavated and piled along the borders. We think that it is significant that these Animas phase sites have ball courts but we should remember that they could be constructed and maintained with little investment of time or labor. Ten people working half a day could probably construct these courts once the area had been cleared of vegetation.

Whalen and Minnis (1996) identified a total of 21 ball courts in the Casas Grandes interaction sphere, and the location of the Culberson ball court brings the total number of courts to 22. Although the courts in the sample are similar in many ways, Whalen and Minnis (1996) also note that there is a tremendous amount of variability. Our study supports this observation. Joyce Well, Culberson, and Timberlake ball courts are located within easy walking distance, they occur at large and apparently contemporaneous pueblos yet there is a significant amount of design variability. Joyce Well has a central depression and berm with double-rock walls and a crude I shape on the southern end. This I shape is not to be confused with the more formal type I-shaped ball courts in Chihuahua, but this is the first time that such features have been identified on the ends of the open-court varieties. Timberlake does not have a depression but it does have court-end rock features and only a single-rock row pattern. Culberson is even a smaller and more simple court. Thus, even within closely associated people, there is a significant amount of court variation.

Ritual Performance

We can never know for certain exactly what went on in these courts, but based upon the attributes of the features, their context, ethnographic and archaeological descriptions of court function in Mesoamerica, and by conducting a performance-based analysis of the feature, we can make some conservative inferences about court use behavior. The ball court's design consists of a set of technical choices, such as the size, shape, orientation, and relationship to the pueblo, that are selected based upon the feature's performance in activities during its life history. In the following section, we outline the performance characteristics and associated technical choices important to Animas Phase ball courts in manufacture and use.

Manufacturing Performance

The excavation and analysis of the ball court at Joyce Well clearly illustrates that two performance characteristics were heavily weighted by the constructors of the ball court: ease of manufacture and ease of maintenance.

1. *Ease of Manufacture*. Because Joyce Well has a slight central depression it would have taken longer to manufacture than Culberson or Timberlake. Nonetheless, the Joyce Well ball court would have been quite easy to

manufacture. A small group of people could excavate the central depression in a short time, and the simple rock alignments could be accomplished rapidly once the orientation had been determined. The floor of the ball court is simply hard-packed sediment, no plaster was applied. It would have been relatively easy to make a flat playing surface. Finally, the rocks used in wall manufacture are readily available in Deer Creek, located just 100 m to the south. The rocks are of the size to be easily carried by a single individual.

2. *Ease of Maintenance*. The Joyce Well ball court would also be easy to maintain. Without a formal plastered floor, the court surface could be maintained simply by sweeping it clear of debris and sediment that may have come into the court between playing sessions. Likewise, the walls of the court would require little to no investment of labor. By simply righting a rock or some other equally simple task the walls could be easily maintained.

Contrast this with ball court 2 at Casas Grandes, which is deeply excavated and has formal vertical stonewalls. Such a court would require a far greater investment of time and labor. Ease of manufacture would not be heavily weighted by the constructors of this court. Likewise, this court would take far more effort to maintain.

Use Performance

The three performance characteristics important in use are accessibility, capacity, and visual performance.

1. *Accessibility*. The ball court is located outside of the pueblo and would be accessible to all villagers. Without a high wall or other impediments, large numbers of people could ring the court and observe the performance. The circumference of the Joyce Well court is 117 m. If you allow 1 m per person, 117 people could ring the court standing shoulder to shoulder. Certainly 200 plus people (probably all members of the community) could be accommodated easily to observe the action if people were to sit and stand. Compare this to dances and other activities performed inside the plazas. Joyce Well has at least two plaza groups, and performances conducted in each plaza would be bounded by the surrounding houses and access to the rituals could be controlled. There are, however, no natural or created impediments to viewing the ball court activities. All people, regardless of social group or standing, could witness the performance. High accessibility, therefore, is a highly weighted performance characteristic among the Amimas Phase ball courts. This is not true in some Mesoamerican ball courts where ball courts sometimes are located in central courtyards or other areas where access is more restricted (e.g., Kelley 1991; Kowalewski et al. 1991; Santley et al. 1991). In terms of accessibility, these ball courts are more like the plazas at Joyce Well and not the ball courts.
2. *Capacity*. Animas Phase ball courts are almost as wide as long. This is a design that would permit many people to participate in the court activity,

which contrasts with some courts from Mesoamerica that may be only 3 m wide (Kowalewski et al. 1991:28–29). Ethnohistoric and iconographic information suggests that the later courts were used in games with one or two players per side. These courts are more like a long racquet ball court, whereas the Animas Phase courts are almost as big as half of an American football field. Certainly, one or two people could play in these courts, but the size of the Joyce Well, Timberlake, and Culberson courts would accommodate large number of people. This is a court that is designed for a game to be watched and played by many.

3. *Visual Performance*. The constructors of the court had many choices in design related to visual performance, which are a suite of performance characteristics that speak directly about the types of rituals performed. What we are talking about here is visual concordance. In order for this feature to play a role in a ritual performance for the Animas people, it must possess a set of attributes (technical choices). Regrettably, we do not have eye-witness accounts of the ball court performances or paintings that represent some aspect of the game as they do in Mesoamerica, but we do have these technical choices as tangible traces of the rituals performed. These technical choices and performance characteristics of the feature itself are important components of the bundle of traits that make up ritual performance.

Given the great design variability between courts, the most striking visual performance characteristic of the Animas Phase courts is orientation. The long axis of all three courts is oriented within just 6° of true north. Of the 15 Chihuahuan courts recorded by Whalen and Minnis (1996:738), 13 were approximately oriented true north. What is more, two of the three ball courts at Casas Grandes are oriented true north (Court 3 at Casas Grandes is the exception but this court is also unique in that it is built right into the room block (Di Peso et al. 1974: 618–620)), as is the Joyce Well room block walls. This concern with cardinal direction and north in particular is clearly manifest at Casas Grandes as the walls of the pueblo are oriented in the cardinal directions, and the Mound of the Cross has four axes that point north, south, east, and west, the so-called "cardinal direction datum" (Di Peso 1974:409). Lekson (1999) has taken the notion of ritual obsession with cardinal directions to the extreme and suggests that the important sites of Chaco Canyon, Aztec, and Casas Grandes were occupied sequentially along the "Chaco Meridan," which follows a latitudinal line. Now is not the time to go into the merits of the meridian argument, but Lekson (1999) does clearly demonstrate that many people of the late prehistoric southwest did orient their architecture in cardinal directions and that north had special significance.

This concern for cardinality is also found with ball courts outside of the Casas Grandes region. The Hohokam built 207 ball courts between AD 700 and AD 1250, and there is a tendency for the features to be oriented either north–south or east–west (Wilcox 1991; Wilcox and Sternberg 1983). This pattern of ball court orientation extends into the Mesoamerican region as well. For example, Kowalewski et al. (1991) found that prehispanic ball courts were oriented either in a near true north–south or east–west direction.

The emphasis on north and the importance of cardinal directions also can be found in the historic period (e.g., Fewkes 1892; Ortiz 1972; Parsons 1996). The true north orientation of the Animas Phase courts is part of their ritual performance that links them, generally, to ceremonies that are significant throughout much of the Southwest and into Mesoamerica as well.

Primary Functions of Animas Phase Courts

Based upon these performance characteristics and other contextual data we argue that there are three primary functions of the Joyce Well, Timberlake, and Culberson ball courts: community integration, celestial-based fertility rituals, and integration in the Casas Grandes religious interaction sphere.

Community Integration

We think that it is significant that the Animas Phase ball courts are located at the edge of the pueblo and not built within it. Both within-community and outside-community ball courts are found in Mesoamerica. In each case, Gillespie (1991) argues that the ball courts function to maintain boundaries in the society. In the cases where the courts are built outside on the edge of the community, she suggests the primary function of the ball court was to symbolize the "segmentation of, and maintenance of the 'correct' distance between, sociopolitical categories" (Gillespie 1991:343). This may have been an especially important function for the Animas Phase courts because sites like Joyce Well, Culberson, and Timberlake represent the largest communities in the region before or since. This is a period of aggregation in the region that brought together people who had previously been living in their own separate communities. It is likely that each plaza group at Joyce Well is composed of interrelated individuals who moved to the village as a group. In this context, the high-accessibility, high-capacity ball court built outside of the pueblo would have provided a means for an entire village to participate. Ball court games would then serve as a means to integrate the social groups in the village.

Celestial-Based Fertility Rituals

After a review of all Mesoamerican ball games and, in fact, many pan-American games, Gillespie (1991) argues that there is a unifying, underlying function. Based primarily on iconographic representations and postcontact writings, she argues that there are two related themes to ball court games (Gillespie 1991:318–321). The first is the symbolic reenactment of the "struggle between day and night, between light and darkness" (Gillespie 1991:319). The games, therefore, symbolize the cyclical journey of the sun, moon, and other "celestial bodies." Although the details for these descriptions and relationships come from far to the south (e.g., *Codex Colombino*, and *Popol Vuh*), we argue that the Animas Phase ball courts shared this basic symbolic theme because of its north orientation, which is consistently associated with celestial movements. Wilcox and Sternberg (1983; see also Wilcox 1991) make a similar claim for Hohokam ball courts. They argue that, "Court orientation may then have been keyed to an annual progression of calendrical ceremonies designed to keep the universe moving smoothly through its annual cycle" (Wilcox and Sternberg 1983:212).

The second but related theme of the ball courts suggested by Gillespie (1991) is agricultural fertility (see also Wilkerson 1991). A number of scholars have suggested that the game was played to ensure the continuation of the cycles of the moon and the sun, which are essential to agricultural fertility (e.g., Parsons 1996). An important component of these games was the real or symbolic sacrifice of a player that symbolized the death and rebirth of the sun and the moon.

Integration in the Casas Grandes Interaction Sphere

Similarities between Joyce Well and Casas Grandes ball court design and other material culture are unambiguous. The people of Joyce Well had choices regarding the type of pottery, hearth design, and architectural style but they chose to mimic Casas Grandes. The rudimentary I shape of the Joyce Well court, in particular, links it to Casas Grandes and places it within the interaction sphere. These similarities are striking and we argue that they are part of the religious interaction sphere centered at Casas.

Whalen and Minnis (1996, 1999) argue convincingly that only the communities in the central zone (within a days walk) around Casas Grandes participated in the exchange of prestige goods. Our three sites were not under the direct political and economic control, as described by Di Peso (1974) for the entire Casas Grandes system. Instead, we argue that the pattern of behavior is most likely understood in terms of religious interaction and pilgrimage (see also Fish and Fish 1999).

Religions are often described as ideologies or systems of belief, but their organization also entails concrete interactions between people and artifacts that have practical goals and result in tangible traces in the archaeological record. Adobe pueblos enclosing a plaza, locally made Ramos Polychrome and other Chihuahuan Polychromes, and ball courts represent some of these tangible clues that the people of the boot heel were participating in this regional system. Ethnographically, in middle range societies, local and regional cults often organize religious activities including household, community, and pilgrimage activities. Crown (1994) has argued that the spread of Salado Polychrome is best explained by appearance of a regional cult near the end of the thirteenth century. We would argue that a similar phenomenon is happening during the same time period in the boot heel of New Mexico and the rest of the Casas Grandes interaction sphere. One difference between the Southwest Regional Cult (and for that matter the Kachina Cult) and what we call the Casas Grandes Ritual Interaction Sphere is that the latter has a definite central place, a pilgrimage center.

Conclusions

This chapter demonstrates how the model can be used to infer religious and ritual performance in prehistory. We also reported on the first excavation of an Animas Phase ball court and the discovery of a third ball court (Culberson) in the boot heel. These are simple features made by placing two parallel rows of rocks and sometimes removing soil from the interior and piling it along the walls. Nonetheless, our performance-based analysis of the feature suggests that the ball courts played an important role in community integration and in placing the sites within the Casas Grandes ritual interaction sphere. The high capacity and high accessibility, and its location away from the pueblo suggest that the game played an important role in community integration and in the maintenance of social boundaries. The truth north orientation of the courts links it to the celestial based fertility rituals and the site of Paquimé located 90 km to the south.

We are just beginning to understand the nature of the Casas Grandes Ritual Interaction Sphere and the role of the Animas phase sites. Research in the coming years will focus on understanding community and household religious organization at Joyce Well and other boot heel sites, exploring the evidence for pilgrimage behavior, and finally understanding the relationship between these Animas Phase sites and Casas Grandes.

Chapter 7
Social Theory and History in Behavioral Archaeology: Gender, Social Class, and the Demise of the Early Electric Car

Archaeology today, it is well known, lacks a unified theoretical framework. Two traditional paradigms – culture history, with its diffusionist theory, and new (or processual) archaeology, with its weak amalgam of neoevolutionary, ecological, and systems theory – have long dominated the discipline's social theory, that is, the principles that explain variability and change in human behavior (Schiffer 1988b). Since the early 1960s, however, three additional theoretical frameworks have arisen, in part as reactions to the many shortcomings evident in the conceptual structure of new archaeology. Behavioral (e.g., LaMotta and Schiffer 2001; Reid et al. 1975; Schiffer 1976, 1992, 1995a; Schiffer and Miller 1999a; Skibo et al. 1995), evolutionary (e.g., Dunnell 1978, 1980; Hart and Terrell 2002; Hurt and Rakita 2001; O'Brien 2005; O'Brien and Lyman, 2002, 2003b; Teltser 1995), and postprocessual archaeology (e.g., Hodder 1985; McGuire 1992; Shanks and Tilley 1997) are minority programs whose advocates seek a wider following. In both evolutionary and postprocessual archaeologies, the major products are historical narratives. Behavioral Archaeology, however, strives to generate both historical narratives and general principles.

This chapter enters the arena of dispute with evolutionary and postprocessual archaeologies by presenting a case study in Behavioral Archaeology. The purpose is to showcase a behavioralist approach to building social theory and to constructing historical narratives. In Behavioral Archaeology, there are intimate and mutually reinforcing relationships between science and history.

While evolutionary and postprocessual archaeologists were seeking to establish hegemonies over social theory in archaeology, behavioral archaeologists were creating a sound basis for inference. It had been shown that underlying every inference are law-like statements that, along with other kinds of information, link observations on the archaeological record to behaviors of the past (e.g., Schiffer 1972, 1976). Regrettably, the principles required for behavioral inference were underdeveloped and unsophisticated. Thus, Behavioral Archaeology's highest priority, when it emerged in the early 1970s at the University of Arizona, was to improve inference by promoting a better nomothetic understanding of material culture and of the formation processes – cultural and noncultural – of the archaeological record.

Since the 1970s, countless studies – many experimental and ethnoarchaeological – have furnished a host of basic principles (correlates, c-transforms, and n-transforms). With contributions from innumerable investigators, this component of Behavioral Archaeology's program is clearly coming to fruition: behavioral inference is now on a firmer and ever-improving footing. One unexpected by-product of this success has been the near-exclusive identification of Behavioral Archaeology with formation-process studies, experimentation, and ethnoarchaeology. Indeed, most archaeologists seem unaware that behavioralists also build social theory and craft historical narratives.

Behavioral Archaeology and Social Theory

Although the first sentence of the book *Behavioral Archeology* promises "a work in archeological methodology" (Schiffer 1976:ix), the remainder of that paragraph hints at a greater program, one that can rise, eventually, on a base of sound behavioral inference:

> If it [*Behavioral Archeology*] is consulted in search of ready-made explanations for the more popular issues in archeology—e.g., the adoption of agriculture, development of civilization, and Mousterian variability—the reader will surely be disappointed. If, on the other hand, the reader is concerned to ask these important questions in new ways and to devise more appropriate strategies for answering them, then this book may be of some interest. (Schiffer 1976:ix)

Readers who advanced to the second page of Chap. 1 would have discovered that new archaeology's social theory was a scant improvement over that of culture history:

> We simply have substituted one set of all-purpose causes—population pressure, environmental change and stress, various forms of intercultural contact, and assorted cybernetic processes—for an equally inadequate set of predecessor causes, such as innovation, diffusion, and migration. At the level of explaining behavioral and organizational variability and change, much of the new has not surpassed the old. (Schiffer 1976:2)

According to behavioral archaeologists, new and better social theory would be developed as attention focused on relationships between human behavior and material culture. These relationships were believed to capture the core concern of the discipline and the most distinctive characteristic of human societies. Studies of behavior–artifact interactions in all times and all places would furnish a framework of principles, new to the social and behavioral sciences, for understanding variability and change in human behavior. Indeed, behavioralists tended to eschew the adoption of social theory from outside the discipline, contending that archaeology's unique focus on long-term behavior–artifact interactions was the only sound basis for generating social theory in all of science (Rathje and Schiffer 1982; Reid et al. 1975; Schiffer 1975a).

Just as Galileo's telescope revealed, literally, a new universe of phenomena for astronomers to explain, so too would an emphasis on people–artifact interactions change the phenomenological world of behavioral scientists. By privileging the study of human activities, the nexus of such interactions, archaeologists would show that behavioral science could not be behavioral or scientific unless it also attended to artifacts. Creative descriptions of this previously unperceived reality would supply the key to constructing new social theory.

Efforts to achieve these ambitious goals were properly subordinated to forging the tools required for reconstructing a behavioral past. Even so, the last chapter of *Behavioral Archeology* contained one example of the creation of social theory. Not only was a rudimentary model presented for status-symbol distribution and change, but also the model's implications were elaborated for explaining certain classes of technological change in complex societies having high social mobility. More recent attempts on the part of behavioralists to build social theory can be found in several chapters and papers (e.g., McGuire and Schiffer 1983; Schiffer 1979, 1992, 2000, 2005a; Schiffer and Skibo 1987, 1997; Walker and Schiffer 2006; Zedeño 1997), a monograph (Schiffer 1992), and a textbook (Rathje and Schiffer 1982). A great many other studies, carried out by investigators who do not identify themselves as behavioral archaeologists, also fall within the scope and spirit of the behavioral program and have contributed important new principles. The present chapter obviously cannot present even a small sample of Behavioral Archaeology's social theory. It is convenient to term the latter "behavioral theory" to distinguish it from the products of other programs.

Behavioral Archaeology and History

New archaeologists and behavioral archaeologists, some believe, are hostile to history. With respect to the latter program in particular, this belief is unwarranted. Indeed, of the four original strategies of Behavioral Archaeology, I and IV are idiographic – that is, historical (Reid et al. 1975). In strategy I, for example, the investigator uses "material culture that was made in the past to answer specific questions about past human behavior" (Reid et al. 1975:864). Such questions, which can be descriptive or explanatory, are quintessentially historical.

Behavioral archaeologists argued that archaeology had both nomothetic and idiographic strategies and that the discipline's vitality depended upon their interdependence (Reid et al. 1975:867). To wit,

> Archaeology can enjoy hybrid vigor by nurturing both its historical and behavioral science roots. Archaeology will be history as long as historical questions continue to be asked. Archaeology will be behavioral science as long as the answering of historical *and other* questions leads the archaeologist to invent and test nomothetic statements in domains that have not been appreciably explored by other behavioral scientists. (Schiffer 1975a:844, emphasis in original)

Clearly, even before the publication of *Behavioral Archeology*, behavioralists had resolved the incipient science–history split that threatened to sunder the discipline.

What is more, behavioralists have endeavored to answer historical questions. For example, among Schiffer's works are prehistoric studies in northeastern Arkansas

(Schiffer and House 1975) and southwestern Arizona (McGuire and Schiffer 1982) and research on electrical and electronic technologies (Schiffer 1991, 2005b; Schiffer, Butts, and Grimm 1994; Schiffer et al. 2003). The latter studies have generally been well received by historians of technology, suggesting that the behavioral approach lays a suitable foundation for constructing narratives.

Although doing history is important in Behavioral Archaeology, so far we have failed to specify, in general terms, the character of a specifically behavioral history. A few words on that subject are now appropriate.

As the name implies, a *behavioral* history is one that strives to explain changes in behavior – that is, alterations in concrete activities. This means that the first concern in most studies is to infer, rigorously, the activities of interest.

Activities fundamentally involve the patterned interaction of people and artifacts. There are many kinds of people–artifact relationships in activities. Indeed, the concept of "relationship" is deliberately left broad and open ended in order not to exclude promising avenues of inquiry. To facilitate communication, however, we have identified several fundamental relationships based upon an artifact's contributions to the activity. These are known as techno-, socio-, and ideo-functions (Rathje and Schiffer 1982:65–67; Schiffer 1992:9–12).

A techno-function is a utilitarian function: the containment, manipulation, or alteration of materials. A socio-function involves the communication of information about social phenomena among an activity's participants or between that social unit and others, so as to affect interaction and activity performance. Artifacts with socio-functions also establish socially appropriate settings for carrying out specific activities (for a more detailed discussion of socio-function, see Schiffer 1992:132–133). When an artifact encodes or symbolizes ideas, values, knowledge, and so forth, it is said to be serving an ideo-function; clearly artifacts with ideo-functions also influence social interaction and activity performance.

The artifacts (and people) taking part in specific activities have, by virtue of their material composition and form, various properties that affect their suitability for interacting in specific ways. These activity-specific capabilities are known as performance characteristics (see also Chap. 1; LaMotta and Schiffer 2001; Schiffer and Miller 1999a, Chap. 2; Schiffer and Skibo 1987, 1997; Skibo and Schiffer 2001) and can pertain to techno-, socio-, and ideo-functions. Factors such as initial cost, maintenance cost, and replacement cost play a role in defining activity-specific relationships between people and artifacts and can, in principle, be treated as performance characteristics (Schiffer 2005b).

Many kinds of behavioral histories, faithful above all to people–artifact interactions, are conceivable. One such general approach, outlined here, focuses on the activities in which a specific artifact participated. The first question is, what are the relevant activities? In some cases, the investigator may take an interest in the reference artifact's entire life history/behavioral chain, including processes of material procurement, manufacture, distribution, use, maintenance, and disposal. The activities making up each process can be characterized in terms of their constituent artifacts as well as specific people–artifact interactions. Also asked of each activity is, what are the techno-, socio-, and ideo-functions of the reference artifact and other relevant artifacts? And which performance characteristics enable the artifacts to carry out those functions?

Next, one can turn to the social units of activity performance. How was each group constituted? Group membership may be defined on the basis of specific performance characteristics – for example, strong people or people with certain kinds of knowledge – or particular variables such as age, sex, gender, occupation, wealth, and kin or corporate or residential group membership. Once the composition of the task group is given, one may investigate its activity-specific ideology (i.e., attitudes, values, beliefs).

These basic data serve as a foundation for examining a host of other relationships, especially the dependency relationships that link any of the reference artifact's activities to other activities. *Dependency relationship* refers to the manner in which two activities are coupled to one another through material flows (inputs and outputs). The study of dependency relationships allows one to trace causes and consequences of activity change (Schiffer 1979, 1992, Chap. 4).

Another tack is to compare the reference artifact to other artifacts having similar functions. One could ask, for example, what are the similarities and differences in performance characteristics between the reference artifact and possible alternatives in specific activities?

To illustrate the character of a behavioral history constructed according to the foregoing model, the case of the early electric automobile in the United States is explored. This example is intended to demonstrate that behavioralists can generate deeply contextualized, engaging, and instructive narratives capable of reaching an audience of nonspecialists – even the general public. For present purposes, the story that follows has been highly condensed (from Schiffer, Butts, and Grimm 1994). To keep it uncluttered, neither references nor justifications of inferences are included. Finally, as a further nod to economy of expression, it is left to the reader to imagine how the behavioral narrative that follows would differ from those that might be fashioned by evolutionary archaeologists, postprocessual archaeologists, or historians.

The Narrative: What Happened to the Early Electric Car?

After many decades of experimentation with self-propelled road vehicles, American inventors and entrepreneurs began to bring their creations to market in 1895. A few years later, in 1900, automobiles powered by steam, electricity, and gasoline competed on a more or less equal footing. Many knowledgeable observers believed that each kind of vehicle would find its own "sphere of action" and that all would coexist indefinitely. In the end, though, the gasoline-powered motor car conquered the others with stunning speed and thoroughness. The electric car's market share declined from 28% in 1900 to less than 1% in 1915. By 1920, the electric car as a commercial product was nearly dead.

Why did the electric car, in contrast to the gasoline car, fail to reach middle-class Americans? An appreciation for the performance characteristics of the two kinds of automobiles in relation to the specific activities of specific groups of people (defined by class, gender, occupation, and rural or urban residence) can help us understand why the electric car failed to find more than a minuscule market.

In 1901, commercial interest peaked, with 41 firms selling an amazing variety of electric vehicles. Like most gasoline and steam cars at the turn of the century, electrics were expensive, ranging from $1,000 to $5,000 at a time when a common laborer might earn $500 a year.

During these early years consumers experimented with automobiles, trying them out in various traditional leisure and practical activities. People of means, mostly men, tested gasoline, steam, and electric cars as replacements for horses and bicycles in racing, for horse-drawn carriages and wagons in trips around the town, and between farm and city, and, most importantly, for bicycles and trains in long-distance touring in the country. The performance characteristics of each kind of car were assessed in relation to these activities. Farmers, who sometimes lived far from town and almost universally lacked electricity at home, found quickly that gasoline cars were a better substitute for horse-drawn wagons than electric cars. Similarly, wealthy urban men discovered in short order that the electric car's limited range on one charge of the battery (20–40 miles), long recharging time (6–10 h), and low speed (12–18 mph) made touring difficult. As automobilist Henry Sutphen (1901:197) bluntly asserted, "Electricity is manifestly out of the question for touring."

Although gasoline cars were unreliable, dirty, smelly, hard to start, and expensive to operate, wealthy automobilists turned to them almost exclusively as the activity of touring became the *sine qua non* of automobilism in the first years of the new century. Automobile magazines, written by and for enthusiasts, as well as mass-circulation magazines, glamorized endurance runs and tours, elevating the mostly male adventurers into heroes of the day – people whose activities would be worthy of emulation by members of the middle class.

Touring cars were built rugged for rough country roads, had engines of four or six cylinders that were powerful for the time, and could go fast – already 40–60 mph by 1910. Significantly, the tourist did not have to worry about where to buy gasoline on the road; having a number of mundane uses, it was available at any country store.

Clearly, the design of the gasoline car was dictated by the touring function, its form and performance characteristics tailored to the leisure activities of elite men. At $1,500–$5,000, the open-air touring car was a bit pricey for most middle-class Americans. Even so, sales of touring cars surged, and a few entrepreneurs – Henry Ford most prominently among them – improved their reliability and repairability and brought down their price. In 1908, Ford introduced the Model T at $850. Within a few years, as the Model T's price dropped, the middle class embraced the gasoline-touring car in large numbers.

Although farmers and male automobilists scorned the electric car, it did find some satisfied customers. Women in particular – all well-to-do, of course – immediately took to electric cars because they were clean, quiet, reliable, easy to start, and simple to operate. In addition to these performance characteristics, the closed-coach styles of the increasingly popular coupe and brougham could be driven in rain, snow, and cold weather. Significantly, the electric car's speed and range were adequate for the urban woman's everyday activities, such as running errands and socializing. The regal electric car was a perfect replacement for the horse-drawn carriage for travel in town and could even carry out the carriage's social functions. No longer dependent on carriage drivers, the wealthy woman in her electric car enjoyed unprecedented independence and mobility.

Appreciating that the performance characteristics of the electric car made it the vehicle of choice for getting around town, even some men, such as salesmen and doctors, adopted it for use in their professional activities. To enhance the electric car's appeal to men, manufacturers began to offer a "roadster" body style that mimicked – in looks only – the stereotypical gasoline touring car.

Performance characteristics of electric cars improved greatly between 1900 and 1910. The use-life and energy density (stored power per pound) of batteries advanced almost yearly, and carmakers reduced energy-wasting friction in the drive train. The happy result was that by 1910 electric cars could travel 50–100 miles on a charge. Owing to the low speed limits geared to the pace of horse-drawn vehicles (usually 12 mph or less), an electric car on one charge could cruise the city all day long.

Unfortunately, the recharging of batteries in 1910 could be difficult, because fewer than 10% of city residences were wired. Outside the home a variety of garages, patterned on livery stables, sprang up to charge and care for electric cars and deliver them to their wealthy owners. The usual stabling fee was $25–$40 per month – about what a working-class person earned. Outside of cities, getting a charge was nearly impossible.

Realizing that charging of car batteries could become a significant source of income, in 1909 the larger electric companies joined carmakers in a promotional campaign. They believed that the electric car, a perfected technology, was poised to take off, even though its market share was now less than 5%.

During the electric car's classic age (about 1910–1914), advertising exploded across the pages of newspapers and magazines. Gradually, discussions of mechanical and electrical virtues – aimed mainly at men – took a backseat to the promotion of comfort, convenience, and luxuriousness. In highlighting these performance characteristics, carmakers were targeting women, whom the ads depicted extensively. In electric car ads published in *Literary Digest*, for example, images of women outnumbered images of men in the ratio of three to one, and women were shown more often as drivers, sometimes chauffeuring men. In one fascinating Detroit Electric ad of 1912, a lone woman heads to her electric car carrying a set of golf clubs. Clearly, the all-weather, easy-to-drive electric car made it possible for wealthy women to enjoy, during the day, a liberated lifestyle.

In the evenings, the electric car became the elegant town car, taking elite couples to the opera, concerts, and the theater. An electric coupe or brougham, with its plush upholstery, curtains, and polished brass or silver fixtures, enabled members of America's horsey set to travel around town in a style once reserved for European royalty and to communicate their exalted social position to friends, acquaintances, and onlookers.

Although sales of electric cars accelerated in the early teens – around 6,000 were sold in 1912 by at least 20 manufacturers – their market share continued to decline. That same year Ford alone produced 82,388 Model Ts, which sold for as little as $525 (compared with $850–$5,000 for an electric car).

For the urban elite seeking to replace a horse and carriage for evening travel in town, the electric car was the motor car of choice. After all, who would want to crank-start a gasoline engine while wearing a tuxedo or gown? Beginning in 1912, however, the horsey set had an alternative. In that year Cadillac, which for some years had already been copying the electric car's closed-coach style, brought out a gasoline town car with an electric starter.

In the next few years, sales of Cadillacs and their clones began to cut deeply into the electric car's core market. Electric car sales stagnated at 6,000 in 1913, and in 1914 began to slump. From 1915 to 1920, the electric car faded into obscurity as, one after another, manufacturers of electric cars went out of business.

It would be easy to conclude that the rapid adoption of the electrified gasoline town car killed off the electric car. Although partly true, that explanation does not account for the electric car's failure to be adopted by the middle class. While inexpensive gasoline cars – still crank started – were finding a huge middle-class market, inexpensive electrics (under $1,000) of the mid-teens were being largely shunned by consumers. A prosperous middle-class urban family could have afforded a cheap electric car but instead chose a gasoline car – even though the price of gasoline was rising while that of electricity was falling, and millions of middle-class homes were now wired. The reasons for this choice are fascinating.

Doubtless, both men and women of the middle class longed to own cars, to emulate the activities of the wealthy. For a woman, an electric car was the ideal city car that could give her, during the daytime, a freedom of action impossible with trolleys. And, of course, it was a car designed for feminine tastes that increasingly were being molded by mass-circulation magazines. As *Electric Vehicles* put it in September 1916 (p. 98), "There is hardly a woman living who would not like an electric." The middle-class man, on the other hand, mainly coveted the car that promised to make possible the adventure and excitement of touring. In ads everywhere and on city streets he could see that the real man's car was a gasoline touring car like the Model T. An electric roadster may have looked like a touring car, but everyone knew it did not perform like one.

In very wealthy families, the conflict over cars was easily resolved by buying two. Many of America's elite, like Thomas and Mina Edison and Henry and Clara Ford, owned "his-and-her" automobiles: one a gasoline touring car, the other an electric coupe or brougham.

Middle-class families lacked the wealth to buy and maintain two cars, and so the decision about which one to buy became a struggle. Most likely, the husband was able to convince his wife that a gasoline car could do more than an electric car and was cheaper too, and thus it was the only sensible purchase. A wife unswayed by this argument could always be reminded that the husband was entitled to make the decision because he was the family's breadwinner. At this time, married middle-class women did not work outside the home. In any event, the struggle between the sexes in middle-class families ended with the purchase of a gasoline car. Had such families been wealthier or had middle-class women enjoyed greater economic independence, the electric car in the teens might have found a market of millions.

History and Social Theory: The Electric Car Revisited

Although behavioral archaeologists can and do fashion historical narratives, even those suitable for the general public, idiographic research is also a source of *nomothetic questions* that can orient theory building. This is an example of the interaction and integration of strategy I with strategies II or III (Reid et al. 1975). In this section, the history of the electric car is used as a springboard for developing a new behavioral theory.

All historical narratives achieve plausibility because the writer and the reader hold in common particular theory- or law-like generalizations (Spaulding 1968). These generalizations connect the causative factors enumerated in the narrative to the event or process to be explained. In most historical narratives, however, the principles – which may be little more than folk theory or ideology – are deeply embedded, invisible on the surface. That theories and laws are implicit is an unavoidable consequence of the narrative form; a story constantly interrupted by exegeses of general principles would be choppy and dull. In a scientific context, however, bringing to light the hidden nomothetic apparatus is essential. Such an exercise may lead to generalizations of potentially widespread applicability. Once explicit, these theory- and law-like propositions can be evaluated for their fit with other principles as well as subjected to testing on new historical cases.

The electric car narrative contains much implicit behavioral theory, and so a complete analysis here is out of the question. To make the task manageable, the focus is on the end of the scenario: why middle-class Americans failed to adopt the electric car.

In previous chapters (see also McGuire and Schiffer 1983; Schiffer 1992; Schiffer and Skibo 1987, 1997; Skibo and Schiffer 2001), behavioral theory dealing with the compromises entailed in the process of artifact design has been elaborated. It has been shown that owing to the complex linkages between technical choices and performance characteristics, an artifact's design cannot optimize the values of all behaviorally relevant performance characteristics: some are necessarily achieved at lower levels than others. Thus, each artifact embodies compromises in performance characteristics relating, for example, to activities of manufacture, use, and maintenance. The pattern of compromises in each case is determined by behavioral factors pertaining to lifeway and social organization. For example, "high residential mobility favors use of houses that are easy to build but often difficult to maintain. In contrast, greater settlement longevity shifts the balance in favor of more manufacturing effort, which is repaid by houses that are easier to maintain and last longer" (Schiffer and Skibo 1987:600).

In order to lay a foundation for explaining technological variation and change, the investigator constructs a performance matrix. Such matrices allow one to compare the patterns of compromise in the performance characteristics of two or more artifact types. It is now possible to recognize at least four kinds of performance matrices:

1. An *absolute* matrix lists absolute values for all behaviorally relevant performance characteristics.
2. A *relative* matrix indicates which artifact type scores higher on each performance characteristic.
3. A *threshold* matrix specifies, for each performance characteristic, which artifact types exceed a given threshold value (for an example, see Schiffer 2005b).
4. A *weightings* matrix denotes whether or not a performance characteristic was apparently weighted heavily in the design process (for example, see Schiffer and Skibo 1987:607).

Although absolute matrices contain the most detailed information, relative, threshold, and weightings matrices can still reveal major patterns in compromises.

The patterned technological variation systematized in performance matrices becomes the focus of explanation. That final step is taken when one shows, with correlates and other behavioral theory, how specific "factors of lifeway and social organization *condition the acceptability of particular design compromises*" (Schiffer and Skibo 1987:600, emphasis in original).

Originally devised to facilitate explanation of design compromises effected between activities of manufacture, maintenance, and use, performance matrices can be easily modified to allow close study of compromises in use activities alone – as is appropriate for the electric car case. Extended in this way, performance matrices become the tool of choice for investigating the adoption of artifacts by consumers (see also Schiffer 2005b).

In the extended model (a performance matrix that treats use activities exclusively), artifacts may participate in more than one use activity, and in each activity they may have any number of techno-, socio-, and ideo-functions. In specific cases, one begins by identifying relevant use activities and the functions that the reference artifact performs in each. The investigator then enumerates the performance characteristics relevant to each function in each activity. Finally, one constructs a performance matrix.

The theory underlying the extended model rests on the premise that an artifact's performance characteristics cannot all achieve high values in every use activity. Thus, the activities can vary greatly in the degree to which they are performed effectively. For example, a Swiss army knife can be used for diverse activities, from cutting meat to taking apart a radio or opening a beer bottle. However, compared with the unifunctional artifacts that might be employed instead (e.g., butcher knife, screwdriver, and bottle opener), the Swiss army knife does not allow every activity to be performed at maximum effectiveness. Bottles can be opened reasonably well with a Swiss army knife, but it is much less effective in cutting meat and taking apart radios. It follows that in the set of activities that share a multifunctional artifact, there will be compromises in activity performance. The extended model allows the investigators to visualize patterns in these compromises, to see which activities were favored and which were disadvantaged. The focus of explanation becomes the patterned compromises, which have to be linked, through explicit principles, to factors of lifeway and social organization.

In the case of the Swiss army knife, none of its use activities can be performed in the most effective manner. The only use-related activity decisively favored is transport from one activity area to another. Principles of technological organization (e.g., Nelson 1991) permit us to appreciate that this pattern of activity compromises is expectable when there is high user mobility and limited transport capability. As we attempt to explain the patterned compromises in activity performance brought to light by applications of the extended model, new behavioral principles will also doubtless emerge.

With the extended model now in hand, as well as new ways to construct performance matrices, we can return to the electric automobile. In listing the activities in which automobiles were used during 1910–1914, one must grapple with the problem of scale, taking care to strike an appropriate balance between activities narrowly and generally defined. Three activities of use are recognized for purposes of this analysis: touring, running errands in town, and traveling to social functions in town. Table 7.1 shows a threshold performance matrix that enumerates the performance characteristics believed to be relevant to each activity. It should be noted that although each activity has a distinctive set of performance characteristics, some of the characteristics are behaviorally relevant to more than one activity.

Inspection of the performance matrix (Table 7.1) reveals some remarkably strong patterns in the effects of gasoline and electric cars on each of the three activities. Insofar as touring is concerned, the electric car was, so to speak, a non-starter, as the touring impresarios claimed. On the other hand, the electric car was well suited to running errands and traveling to social functions in town. Clearly, neither kind of car allowed all activities to be performed effectively. A household's choice of one car over the other would have been an unhappy compromise that reflected the differential weighting of activities.

Before discussing the problem of how to treat the differential weighting of activities, one must bring the car users into the foreground by introducing the dimensions of gender and class. Although one can generate gender- and class-based performance matrices that present relevant activities and relevant performance characteristics for each abstractly defined user group, the process is simplified here for the sake of brevity. The strength of association between a specific activity and given gender-class groups (its "loading") is discussed.

Touring in the teens was a socially desirable activity for men. It began as a leisure pursuit for members of the upper class, but by 1910 tens of thousands of Model Ts and other inexpensive gasoline cars were being driven by middle-class men striving to emulate the activities of their wealthier brothers. In short order, the middle-class man's social competence – the ability to interact effectively with other men, especially of his class – was coming to depend on the possession of a car capable of touring.

Running errands in town, especially during the day, was an activity with a very high female loading, regardless of social class. What differed by class were the available transport technologies. Upper-class women could depend on horse-drawn

Table 7.1 A threshold performance matrix for gasoline and electric automobiles, circa 1912[a]

Activity	Performance characteristic	Gasoline	Electric
Touring	Range of 100+ miles (T)	+	–
	Top speed of 40–60 mph (T,S)	+	–
	Ease of fueling, recharging (T)	+	–
	Ruggedness (T)	+	–
	Economy of operation and maintenance (T)	–	–
	Repairability in the country (T)	+	–
	Can indicate owner's membership in the group "tourists" (S)	+	–
	Can indicate owner's wealth (S)	+	+
Running errands in town	Range of 50–100 miles (T)	+	+
	Speed of 12–20 mph (T)	+	+
	Ease of starting (T)	–[b]	+
	Ease of driving (T)	–	+
	All-weather capability (T)	–[c]	+[d]
	Reliability (T)	–	+
	Economy of operation and maintenance (T)	–	–
	Ease of fueling, recharging (T)	+	+[e]
	Can indicate owner's wealth (S)	+	+
	Can indicate owner's social position (S)	+	+
Traveling to social functions in town	Range of 50–100 miles (T)	+	+
	Speed of 12–20 mph (T)	+	+
	Ease of starting (T)	–[b]	+
	Ease of driving (T)	–	+
	All-weather capability (T)	–[c]	+[d]
	Reliability (T)	–	+
	Economy of operation and maintenance (T)	–	–
	Ease of fueling, recharging (T)	+	+[e]
	Cleanliness of operation (T)	–	+
	Quietness of operation (T, S)	–	+
	Can indicate owner's membership in the "horsey set" (S)	–	+
	Can indicate owner's wealth (S)	+	+
	Can indicate owner's affinity for "high culture" (I)	–	+

[a] Entries represent an approximation of how these performance characteristics were judged. A plus (+) indicates that the car exceeds the threshold value of that performance characteristic; a minus (–) indicates that the car falls short of the threshold value. T = techno-function; S = socio-function; I = ideo-function.

[b] After 1912, the pricier gasoline cars had an electric starter.

[c] A few expensive gasoline cars, like the Cadillac, had a closed-coach body style, but the touring car exposed the occupants to the elements.

[d] The electric roadster lacked all-weather capability.

[e] In homes without electricity, recharging of batteries could not have been done economically.

carriages and coachmen or on cabs, while middle-class women had to walk or take the trolley. The electric car was quickly adopted by upper-class women as appropriate for running errands.

Men and women both traveled to evening social functions. Again, however, the electric car – the most suitable technology for the activity – was restricted to America's elite; middle-class Americans took cabs or trolleys.

Upper-class households, by buying two cars – gas and electric – were able to avoid making unhappy compromises in automobile-use activities. Middle-class households in the teens may also have desired two cars, for the same reasons, but they simply could not afford their purchase and maintenance costs. Someone's activities had to be severely compromised, and those activities were mainly women's.

Certainly, there is nothing novel about the generalization that wealthy households can afford more artifacts (e.g., Schiffer et al. 1981). What is new is the recognition, grounded in behavioral theory, that wealth makes it possible to avoid compromises in activity performance caused by the employment of multifunctional artifacts. Wealthy households acquire a plethora of artifacts having very narrow functions that enhance the performance of specific activities. This can be stated more formally as the "Imelda Marcos" hypothesis (a pair of shoes for every occasion)[1]: in a class of sedentary behavioral components (e.g., households, corporate task groups, communities), the members with greater wealth are able to enhance the performance of favored activities by acquiring additional specialized or unifunctional artifacts. This means, for example, that a set of techno-, socio-, and ideo-functions formerly performed by one artifact can be carried out by several. In the present context, "unifunctional" does not mean literally only one function; rather, the term denotes artifacts having a reduced or limited number of functions (relative to the artifacts being replaced).

Other processes in addition to that described by the Imelda Marcos hypothesis can also cause unifunctional artifacts to proliferate in specific activities. For example, as Zipf (1949) long ago hypothesized, a tool kit used at a high rate will differentiate into more specialized tools as artisans seek to reduce their effort per unit of output. Zipf's hypothesis is of value in explaining the expansion of tool kits that can accompany changes in the scale of certain production activities, though it appears to apply mainly to techno-functions. The process described by the Imelda Marcos hypothesis, however, operates independently of rates of activity performance and artifact use and covers all artifact functions. Another process is at work when the constraints of high mobility (which favor multifunctional artifacts) are relaxed. As residential mobility decreases, a behavioral component is apt to acquire more artifacts, including those with narrower functions, for some activities. Although neither of these alternative processes is relevant to the automobile case, one should keep them in mind when offering explanations for other instances of unifunctional-artifact proliferation.

The Imelda Marcos hypothesis, though obviously requiring further refinement and empirical evaluation, is not without interesting implications. A few examples should suffice to illustrate its productivity.

In complex societies without rigid sumptuary rules, especially where there is a high social mobility, the acquisitiveness of wealthy households seemingly lacks limits. Entirely new technologies and industries can arise simply to meet the

---

[1]Imelda Marcos, wife of the late Philippine leader Ferdinand Marcos, was fond of footgear. Forced out of power by a coup in the mid-1980s, the Marcos family made a hasty exit from Manila. The first visitors to the abandoned presidential palace discovered Imelda's trove of 3,000 pairs of shoes.

insatiable demands of well-to-do consumers (cf. Schiffer 1976, Chap. 12). It is commonplace to attribute such lavish acquisition behaviors to the ceaseless quest for prestige and high social standing, which is fulfilled – but often only temporarily – by artifacts having appropriate social functions. I suggest that the traditional account is, at the very least, incomplete. It is clear that households that acquire new products at high rates garner greater prestige and social standing in certain activities; the artifacts do serve these socio-functions. But does that effect alone explain the acquisition behavior? The Imelda Marcos hypothesis suggests that another cause may be the effort to enhance the performance of favored activities through the acquisition of innumerable artifacts having narrow functions, including techno-functions. Thus, one should not forget that "status items" and "prestige goods" can also carry out important techno-functions. After all, even Imelda Marcos used some of her shoes for walking. This implication of the Imelda Marcos hypothesis resonates with an analytical imperative of Behavioral Archaeology: artifacts must be deeply contextualized in relation to all relevant activities.

The Imelda Marcos hypothesis also has implications for understanding the use of space. To wit, another strategy for enhancing favored activities is to conduct them in larger and sometimes dedicated – that is, unifunctional – spaces. Thus, when wealthy households proliferate unifunctional artifacts, they may also expand and subdivide their dwellings and tofts.

Because the term *artifact* can, for certain purposes, include people (Schiffer 1979; Schiffer and Miller 1999a), the Imelda Marcos hypothesis is seen to have unexpected utility in accounting for the proliferation of specialists – people who carry out, usually with great skill, a limited number of activities. For example, to enhance the performance of certain activities, elite households may add unifunctional members, such as cooks, butlers, maids, chauffeurs, and gardeners. Doubtless the Imelda Marcos hypothesis can be extended to people in other kinds of behavioral components.

The final implication deals with the effects of unifunctional artifacts on activity performance. The addition of unifunctional artifacts can cause an activity to change in predictable ways. In a word, an enhanced activity is apt to become more differentiated and complex. Owing to the additional artifacts and thus more intricate people–artifact interactions, the spatial organization of the activity also changes; as noted already, often more space – even unifunctional space – is needed. New artifacts require new maintenance activities, which may in turn entail new unifunctional maintenance artifacts and dedicated maintenance areas. These altered material flows establish new dependency relationships between the original activity and others, which can contribute to further activity changes in different parts of the behavioral system (Schiffer 1979, 1992, Chap. 4). In addition, as the task group becomes more practiced at using the new artifacts, tacit knowledge and skill will increase along with the activity's techno-science content (Schiffer and Skibo 1987). The activity's ideology will also change, as the task group adopts more appropriate activity-maintaining values and attitudes (Schiffer 1992, Chap. 7). Clearly, activities enhanced by an infusion of unifunctional artifacts will undergo many changes, having implications for our understanding of how behavioral systems alter in response to the allocation of resources to favored activities.

The Imelda Marcos hypothesis helps us understand how very wealthy households solved the car problem and why that solution was unavailable to the middle class. However, an important question remains: Why did middle-class households favor touring over running errands in town? Another way to ask this question is, why was a heavily male-loaded activity favored over a heavily female-loaded activity? The answer, furnished in the narrative above, implicates the structure of middle-class families. In the traditional Euro-American patriarchal family, men decide which activities are favored, and allocate resources accordingly. Middle-class men, captivated by touring, privileged their own leisure activities, and so bought gasoline cars. This commonsense explanation has some appeal, but in a scientific context it should be the beginning, not the end, of inquiry. What desperately needs investigation is how activities come to be differentially weighted by various kinds of behavioral components.

To facilitate investigation of this issue in the future, one can further generalize the Imelda Marcos hypothesis. Its most fundamental component is that the investment of resources in an activity, to enhance its performance, leads to an increase in unifunctional artifacts. In this fully generalized form, the hypothesis can even be applied to a class of behavioral components having the same wealth. Today, for example, there is enormous variation in the acquisition behaviors of middle-class households (e.g., Schiffer et al. 1981). Much of that variation is likely to be in the form of unifunctional artifacts obtained to enhance the performance of favored activities. What needs explanatory attention, then, is the differential enhancement of activities.

By monitoring the proliferation of unifunctional artifacts, the investigator has a powerful tool for assessing a behavioral component's activity priorities. This perspective permits us to raise old questions about our own society in new ways. For example, why do some lower middle-class households invest heavily in the sport, car-repair, and partying activities of adult men, whereas others channel resources disproportionately into enhancing the educational activities of children? Attempting to create the behavioral principles needed to answer such questions, which is obviously beyond the scope of this chapter, can lay a foundation for much fruitful research on the causes of behavioral variation and change.

Summary and Conclusions

Since the 1970s, behavioralists, along with other investigators, have begun to contribute the principles and procedures needed to put archaeological inference on a scientific foundation. Happily, it is becoming possible to describe some characteristics of past societies in behavioral terms.

Inferring past behavior was never viewed by behavioralists as archaeology's final goal. Rather, behavioral inferences provide the basis for generating a view of the past compatible with a particular theoretical stance: the behavioralist premise that the basis of human societies is their complete reliance on complex and intimate relationships between people and artifacts (Schiffer and Miller 1999a). The study of such relationships, in all times and all places, can, behavioralists maintain, lead to the creation of distinctive social theory in archaeology.

Using the case of the early electric car, it was demonstrated that behavioral theory, immature though it remains, facilitates the fashioning of historical narratives that are both richly contextualized and audience friendly. More significantly, a behavioral narrative is centered on the actual activities of past people.

In Behavioral Archaeology, however, historical narratives are not the only or the ultimate product. For behavioralists, history (i.e., strategy I) can be a source of general questions that serve as a starting point for crafting new behavioral theory (in strategies II and III). The electric car study provided an example of strategy interaction as the narrative was dissected to disclose some of its nomothetic underpinnings. This exercise led to the development of an extended model for studying, with performance matrices, the effects of multifuctional artifacts on activities. This behavioral model allows one to understand the patterns of compromise in activity performance occasioned by instances of product acquisition.

These theoretical discussions, prompted by the electric car case, also led to the formulation of the Imelda Marcos hypothesis, which states that in a class of behavioral components (e.g., households), wealthier members can afford to invest in greater numbers of unifunctional artifacts. The performance of favored activities is thereby enhanced because one can avoid the compromises entailed by the use of multifunctional artifacts. The hypothesis was generalized and additional implications derived. The effort to explain why middle-class households in the teens enhanced male-loaded activities (touring) instead of female-loaded activities (running errands in town) foundered for lack of relevant behavioral theory. Development of the appropriate principles is urgently needed to permit replacement of folk theory and modern ideology, which are, regrettably, the nomothetic basis of many archaeological explanations. The behavioralist demands that historical narratives rest, eventually, on a foundation of well-confirmed behavioral principles. As the consideration of the electric car narrative shows, nomothetic strategies of Behavioral Archaeology can serve history not just by improving behavioral inference but also by answering, with credible theories and laws, the general questions raised in specific narratives. This vision of the mutually beneficial relationship between history and behavioral science in archaeology is an uplifting one, for it encourages individuals to pursue the research activities for which they are best suited and it fosters an across-the-board elevation of standards. Clearly, we can now build behavioral science for distinguishing between rigorous historical narratives and just-so stories.

Although improving historical narratives is a good reason for creating sound behavioral theory, it is not the only reason. Archaeology is also a unique behavioral science that, owing to its emphasis on artifacts, has much to contribute to other behavioral sciences. The foundation of archaeology as behavioral science is "the study of relationships between human behavior and material culture" in all times and all places (Reid et al. 1975:864). Thus, the focus of theory building in archaeology is not on culture or on extrasomatic adaptations or even on the archaeological record but on what people actually do (and did) in specific activities. By privileging people–artifact interactions, behavioral archaeologists are able to discern a distinctive order of human phenomena, previously unperceived, that is amenable to nomothetic study. Constructing behavioral theory to explain variation and change in human behavior, conceived as people–artifact interactions, is archaeology's highest scientific calling.

Chapter 8
Studying Technological Differentiation

A process of large-scale behavioral change commonly encountered in the archaeological and historical records is technological differentiation (Schiffer 1992:107). This process has contributed greatly to technological variation, and so its study should be accorded a high priority. This chapter supplies investigators with the theoretical tools for explaining, in proximate fashion, any technology's differentiation. Such proximate explanations become the foundation for fashioning behaviorally grounded historical narratives.

In the process of technological differentiation, a new technology appears, usually at first in a small number of functional variants. Over decades, centuries, even millennia, that technology becomes diversified as people create and adopt new varieties. As was shown in Chap. 6, Anasazi pottery manufacture began in the first centuries AD initially with a few jar forms, but by AD 1000 jars had been joined by more varied jars, bowls of many sizes, effigy vessels, ladles, and so forth. During the following centuries other variants were adopted, which differed on the basis of shape as well as slip color and painted decoration. By the thirteenth and fourteenth centuries, potters on the Colorado Plateau were making and using dozens of ceramic variants having many utilitarian and symbolic functions. Similar processes of functional differentiation are discernible in other Anasazi technologies, such as ground stone (Adams 1994), ritual artifacts (Walker 1995a), and architecture (Lipe and Hegmon 1989).

Differentiation processes are exhibited by many technologies in societies at every level of complexity. The electronic computer, for example, began its stunning trajectory in the 1940s, when a few machines were made for ballistics calculations. In the 1950s and 1960s, various computers were created for military, industrial, and commercial activities. During the 1980s, entirely new families of computers were developed for business, education, and household use. Today, special-purpose computers can be found in everything from airliners to stuffed animals.

Obviously, Anasazi pottery and electronic computers differ in materials, manufacturing processes, uses, range of symbolic functions, social and cultural contexts, rates of change, and other seemingly salient variables. Nonetheless, we suggest that such diverse examples of technological differentiation have common features, and so can be handled within one theoretical framework – a decided advantage given the enormous range of societies and technologies studied by anthropologists.

For present purposes, *technology* is defined as any type of artifact (e.g., stone axes, cave paintings, electric automobiles), material technology (e.g., ceramics, basketry, silicon), or technological system (e.g., cooking technology, weaving technology, ritual technology, electrical power systems). As is well known, all technologies incorporate knowledge, usually have ideological correlates, and are embedded in social, economic, and political structures. In this chapter, however, these important factors are held in abeyance as the goal is to create a framework for establishing, in behavioral terms, the *proximate* causes of technological differentiation.

There are good theoretical reasons for making this methodological move. In many disciplines today, from history to cultural studies to sociology, research on technology has become acceptable. However, most investigators find themselves ill equipped to deal with the pervasive materiality of human life (on the latter, see Schiffer and Miller 1999a) because they lack the basic conceptual tools needed for describing, and sorting out the proximate causes of, technological variability. Not surprisingly, they force technology to conform to conventional ways of handling other sociocultural phenomena. Thus, technologies are conceptualized immediately in social or cultural terms so that they can be handled as familiar phenomena – cognitive, ideological, social, political, and so on. Often, in such works, one searches in vain for adequate descriptions of the relevant technological variants and the concrete activities in which they were used.

Clearly, various "external" factors are relevant for furnishing fully contextualized explanations of technological variability and change, but I contend that such factors cannot be privileged initially or given methodological priority. Granted that we all seek contextual explanations, our first task is to establish a behavioral foundation for such explanations that highlights interactions between people and artifacts in concrete activities. Once in place, such a foundation enables the investigator to construct deeply contextualized narratives that are far more than just-so stories because they rest on the materiality of human life.[1] In asking new questions about technological variability and change, we need to construct new behaviorally based models and theories, never forgetting that questions about technology are, first and foremost, questions about human behavior (and vice versa).

Despite the revival of interest in technology in anthropology (e.g., Dobres 2000; Dobres and Hoffman 1999; Hayden 1998; Hutchins 1995; Keller and Keller 1996; Lemonnier 2002b; Pfaffenberger 1992; Schiffer 1991, 1992, 2001b; Schiffer et al. 1994a), large-scale processes – those playing out over large spans of time and/or

[1] Some readers may object to the bottom-up treatment of technology transfer; that is, the initial focus on the delineation of communities, activities, variants, and performance characteristics. Suffice it to say that many social-science treatments of technology are poorly grounded in the materiality of human life, a result of the investigator's theoretically mandated haste to immediately engage social, political, ideological, and other factors before establishing a behavioral foundation.

space – have received scant *theoretical* attention (for exceptions, see Gould 2001; Lyman and O'Brien 2000; Schiffer et al. 2001a). Perhaps technological differentiation has attracted little explanatory effort because, on the one hand, for archaeologists and historians it is commonplace, the cause apparently obvious. Yet, the history of science teaches us that the study of ordinary, even mundane, phenomena can lead to the construction of useful theory. On the other hand, such a large-scale process might be essentially invisible to ethnographers if, as is often the case, their units of study are highly constricted in space and time. Regardless of the cause, large-scale technological processes are drastically undertheorized in modern anthropology. In recognition of this lacuna, a simple theoretical framework is put forth, built on the ubiquitous phenomenon of technology transfer, which enables the investigator to study instances of technological differentiation. To illustrate use of this framework, which can help investigators lay a behavioral foundation for erecting deeply contextualized historical narratives, the case study of eighteenth-century electrical technology is offered.

Technology Transfer: A Behavioral Framework

One potentially fruitful tack for understanding technological differentiation is to treat it as a product of processes that come into play when technologies are transferred from community to community within and between societies. By framing the explanatory problem in this manner, the investigator can focus initially on how people carrying out different activities in recipient communities contribute to the redesign and proliferation of adopted variants.

The term *technology transfer* is used with some trepidation because it already sees service both inside and outside the academy. It usually refers to the transfer of a technology from one societal or cultural context to another and is regarded as desirable by some people. As a result, there has been much discussion on how to transfer technologies from military laboratories to the civilian sector, from universities to commercial and industrial corporations, and from more "developed" to "less-developed" regions (e.g., Matkin 1990; Roberts 1988; William and Gibson 1990). Despite its ideological freight, which sometimes justifies the expansion of corporate capitalism, technology transfer remains an empirical phenomenon of great antiquity, for, as the archaeological and historical records demonstrate, technologies have always been transferred within and between societies. However, before anthropologists can employ technology transfer for studying technological differentiation, it should become fully behavioral – that is, conceived at the level of concrete interactions among people and artifacts.

A basic premise of this framework is that technologies are transferred between and among *communities*. This claim, although not novel, has novel implications when community is defined in very general terms. For present purposes, a community – a *techno*-community, if one prefers – is any group of people whose members take part in one or more activities that incorporate variants of a particular technology. Thus, "field archaeologists" comprise a community whose members carry out survey and excavation activities employing an array of familiar recovery and recording technologies. Another community is "archaeological theorists," whose text-creating activities involve writing technologies, the texts of others, and, perhaps, desks and comfortable armchairs. Likewise, one can recognize communities of *bonsai* gardeners, particle physicists, letter carriers, craft potters, taggers, Creole chefs, Internet surfers, rock climbers, philatelists, and Catholic priests, each consisting of a group whose members, regardless of whatever else they do and whether they interact among themselves, conduct certain activities using particular technologies. (The concept of community is elaborated later – see "Acquisition" and "Use" in the following section.)

A second important premise is that, when initially received, transferred technologies often do not perform adequately in the activities of recipient communities, and so must be redesigned. Clearly, the motor of technological differentiation lies in the activities of recipient communities.

Phases of Technology Transfer

Regardless of the nature of the technology or communities involved, technology transfer can be modeled as a six-phase process (see also Schiffer et al. 2003). In the first phase, "information transfer," people learn about a technology or variant through one or more transfer modes, such as word of mouth, written materials, or examples of the technology itself. The second phase, "experimentation," involves a hands-on assessment of the suitability of the new technology for given activities. The third phase, termed "redesign," entails modification of the technology so that its performance characteristics – its behavioral capabilities – become better suited for particular activities of the recipient community. In the fourth phase, "replication," the modified technology is reproduced through one or more replication modes and made available to members of the recipient community (or communities). "Acquisition," the fifth phase, takes place when some people obtain the new (redesigned) variant. Finally, the sixth phase is "use," which is incorporation of the new technology into the recipient community's activities. After people acquire and use replicated examples of the technology, the investigator is able to specify the composition of the adopting community.

These phases, it should be noted, are neither rigidly discrete nor sequential in real-world instances of technology transfer. For example, limited acquisition often occurs before redesign and replication; indeed, one might treat experimentation as an early phase of acquisition. In addition, redesign and replication are usually iterative, coeval processes that can alternate with use. Merely a heuristic device, these phases allow the investigator to break down a complex process into manageable units of study.

Information Transfer

In information transfer, a description of the new technology is conveyed, directly or indirectly, from person to person. Common modes of information transfer, which were all present in the eighteenth century and figured in the transfer of electrical technology, include (1) journals and monographs, (2) trade and text books, (3) newspapers and magazines, (4) meetings of scientific societies and academies, (5) public demonstrations and lectures, (6) teachers in colleges and universities, (7) scientific-instrument shops and catalogs, (8) interpersonal interaction (not subsumed by the above modes, such as corresponding by mail and visits), and (9) nonmarket exchange of the actual technologies through gift, inheritance, and other means.

The precise modes of information transfer, although interesting, are not stressed in the present study. After all, the transfer of information is a necessary but never a sufficient cause of *technology* transfer: most people who learn about a new technology never acquire it. Moreover, because information transfers often leave no material traces in the archaeological and historical records, efforts to infer *in detail* the flow of information among communities may fail. Thus, in the present framework, there is a low priority on ascertaining specific information transfers.

Experimentation

Experimentation usually begins when a few people try out a technology in new activities or forecast in "thought experiments" how it might perform. In this way, knowledge is created about the fit between given activities and the technology's performance characteristics. For example, the first home computers, initially acquired by young electrical hobbyists in the 1970s, were tested by other people in activities such as word processing, record keeping, and game playing; the early experimenters found home computers to be wanting in a great many behaviorally relevant performance characteristics, from ease of use to reliability. Complaints about these performance shortcomings demonstrated that home computers were not well designed for many everyday activities.

Sometimes experiments with the technology, through activities such as play and trial and error, can also indicate its suitability for an entirely new activity. For example, in playing with early bicycles, people showed that this peculiar conveyance could be used for touring in the country. Together, bicycles and touring were widely adopted by a bicycle-tourist community that emerged in the late nineteenth century (Smith 1972).

Redesign

Whether a new technology is tried out in old or new activities, a common outcome is that the technology shows some promise, but its mix, or weighting, of performance characteristics is inappropriate. In response, the technology is modified, or redesigned, to yield a different weighting. For example, portable radios designed originally for home leisure activities were tried out in diverse work activities by policemen, firefighters, and aviators in the 1920s. Such experiments led to the design of several kinds of specialized "mobile" radios; brought to market almost immediately, the new radio variants – ruggedized and adapted to specialized power supplies – were quickly acquired by members of various communities (Schiffer 1991). Likewise, the first shirt-pocket radios, which employed earphones and looked like hearing aids, were shunned by image-conscious consumers; later designs had very different visual performance characteristics (Schiffer 1991).

Technology transfer, then, potentially entails an appreciable amount of redesign and the proliferation of new, and usually specialized, functional variants. This is an important premise because it directs the investigator to seek, in altered activity contexts, the situational factors that might have influenced how the performance characteristics of new variants were weighted. Situational factors include the composition of the activity's social unit, the specific artifacts employed, the precise interactions among people and artifacts in the activity, material flows to and from other activities, and the location and frequency of activity performance (Schiffer and Skibo 1997).

It is helpful to regard situational factors as the proximate causes of a technology's performance requirements. Indeed, it is only by acting on situational factors that potentially relevant "external" variables, such as social inequality, religious ideology, and political structures, can affect a technology's design.

The process of design or redesign – there is no need to distinguish between them in the present framework – has already been discussed in Chap. 1, so only a brief description of that theory is provided as it relates to the present case.

In the design process, the artisan (a gloss for any person or group that designs and manufactures a technology) can receive feedback from other people, such as its sellers and users, whose activities take place along the life history of that technology. From this feedback, the artisan learns about functional requirements, that is, which performance characteristics *ought* to be weighted in particular activities (of marketing, use, maintenance, etc.). Often, the performance requirements of different activities conflict, and so the artisan usually fashions a design – a set of technical choices – that embodies compromises in performance characteristics. By affecting situational factors, innumerable social, political, ideological, and technological variables can influence the *actual* weighting of performance characteristics.

Performance characteristics are a technology's activity-specific and interaction-specific capabilities. Performance characteristics relevant to design run the gamut from those enabling mechanical and thermal interactions to others that come into play when someone views another's clothing, listens to a sermon, or smells dinner cooking. The case study below shows, for example, how several sensory (visual and acoustic) performance characteristics were heavily weighted in the design of some electrical variants having social and ideological functions.

In the absence of experimental data and detailed contextual evidence, performance characteristics can be estimated on the basis of the variant's formal properties (size, shape, etc.) and the functional requirements of its activity-specific interactions with people and other technologies. Even when historical or ethnographic evidence is available, however, assessments of performance characteristics are inferences constructed by the investigator.

Replication

Replication, a term employed by evolutionists (e.g., Leonard and Jones 1987), is the reproduction of a variant – its manufacture and distribution to users. We enumerate three somewhat idealized replication modes (see also Costin 1991).

1. One person makes a new, unique variant of the technology for his or her own use. Such singular technologies, although seemingly ephemeral, can be of great importance, especially in activities of scientific research. What is more, singular technologies are common in all complex societies, often commissioned by elite personages. For example, singular architectural technologies, from palaces to tombs, abound in the archaeological record of ancient states.
2. More than one person constructs copies of the technology for their own use, creating recognizable examples of a particular variant. Many crafts produced by the artisan for his or her own use follow this replication mode, such as cooking pots that are used only at the potters' homes.
3. One or more artisans manufacture a variant and make it available for exchange or purchase. In market-based societies, this replication mode can be referred to as "commercialization" (see Schiffer 1996). Instrument makers in the eighteenth century commercialized many variants of electrical technology.

An appreciation for differences in replication modes helps us to understand how people could have acquired examples of particular variants. In the case study below, all replication modes seem to have been employed by members of all communities; indeed, there were many ways to acquire electrical technology throughout the eighteenth century.

Acquisition

During acquisition, examples of the technology are obtained by people who thereby become members of the recipient community. Community members may consist of all people in an existing community (defined on the basis of another technology or technologies, and related activities), may be part of one or more existing communities, or may form an entirely new grouping.

Communities can be designated at various scales, depending on how the investigator defines activities and technologies. An overarching community of "electricians" is recognized here and is made up of intermediate-scale communities of electrophysicists, electrochemists, electrotherapists, and so on. At a finer scale, one could designate, for example, communities of pneumatic electrochemists and materials-science electrochemists. In this chapter, the focus is on communities at an intermediate scale (e.g., electrochemists and electrotherapists), which is a convenient compromise between coarse- and fine-grained patterns.

Memberships of different communities may overlap to any extent, from completely to not at all. In archaeology, for example, fieldworker and theorist communities share many members; but few archaeological theorists are also *bonsai* gardeners. In the eighteenth century, many people were members of several electrical communities.

Communities can exist within one society or crosscut many. In the case study later, most communities had members from major European polities, especially England, France, Germany, Holland, and Italy. Although the case study has international dimensions, what matters most in the technology-transfer framework is that behaviorally based communities be delineated, whether their memberships cross geographic, social, or political boundaries. As employed here, community is a flexible analytic unit that helps the investigator to formulate questions about technological differentiation.[2]

Acquisition, which we equate for present purposes with "consumption" or "adoption," is much studied in anthropology (e.g., Douglas and Isherwood 1979; Majewski and Schiffer 2001; McCracken 1988; Miller 1987; Spencer-Wood 1987). Commonly, investigators seek to correlate acquisition behavior with cultural and sociodemographic variables. However, from a behavioral perspective, *explanation* of acquisition requires the investigator to estimate and compare the performance characteristics of the acquired technology against alternatives in relevant activities (e.g., Chap. 7; Schiffer 2001a; Skibo 1994). With knowledge of a technology's replication modes, acquisition patterns, communities, community-specific activities, and alternative technologies, the investigator can rigorously pose, and seek answers to, the question, "Why did members of a community adopt technology *y* in preference to technologies *u*, *v*, and *z*?" However, explaining *specific* instances of acquisition goes beyond the immediate explanatory goals of the technology-transfer framework because such explanations are concerned with small-scale, not large-scale, technological change. In dealing with technological differentiation, the investigator need do no more than call attention to the new variants' weightings of performance characteristics, which rendered them more or less appropriate for a recipient community's activities.

Use

The sixth and final phase, "use," comprises the activity or activities in which the acquired technology interacts with community members, when its use-related performance characteristics come into play. By distinguishing *use* from *acquisition*,

[2] Because the activities of a single community can change over time, the investigator may also treat these changes as instances of technology transfer, equivalent to transfers between two different communities. That is how technological changes in the electrophysicist community are handled.

we are able to appreciate that a community can include far more people than just those who *directly* acquired the technology. For example, a household's adults usually obtain food-serving technologies, but these are also used by children and guests in eating and feasting activities. Similarly, school administrators and purchasing agents acquire computerized teaching aids, but once installed in classrooms these technologies are actually used (or not) by instructors and students. The investigator may delineate groups making up a community on the basis of behavioral roles, social roles, age, sex, gender, ethnicity, social class, and other sociodemographic variables. Diversity in a community's membership, and relationships between members and nonmembers, afford the investigator an opportunity to address questions about social power (Walker and Schiffer 2006), access to resources and community membership, conflict and negotiation, and so on. Many technologies – ancient and modern – are adopted by communities having considerable social and behavioral diversity, which affects the actual weighting of performance characteristics and thus the design of new variants (Bijker 1995; Schiffer and Skibo 1997).

Narratives of Technological Differentiation

Although the investigator may employ the technology-transfer framework as a starting point for building new theories of behavioral change and for studying small-scale processes (e.g., the adoption of a particular variant), its most important use is in establishing proximate, behavioral explanations of variability and change. Such explanations, which for each recipient community link situational factors of activities to the performance characteristics of technological variants, serve as a foundation on which to build deeply contextualized historical narratives that implicate external causal factors.

Once the investigator has identified new variants and explained how situational factors affected the weighting of their performance characteristics, he or she has endless possibilities for building historical narratives. Investigators can, for example, fit the details of a case into a biographical framework if a few major personages were involved in transferring a technology and creating its variants. One can also employ a purely chronological framework, focusing on the sequence in which new variants emerged as the technology was transferred from community to community. Yet time can be procrustean, especially when there are many communities, many transfers, and many new variants. In the latter cases, it might be preferable to use the communities themselves as the organizing framework.

Historical narratives also vary greatly in the sorts of details and causal relationships that are emphasized. Some investigators prefer narratives that highlight social agency, from the desires and intentions of individuals to power struggles between genders, social classes, and so on (e.g., Dobres 2000). Others stress the influence of overarching economic and ideological factors. Still others take an evolutionary tack, detailing the comings and goings of particular variants in relation to selective pressures of the environment (e.g., O'Brien and Lyman 2000). Beyond specifying that activities are the immediate source of technological change, the technology-transfer framework – and much other behavioral theory – is agnostic with respect to ultimate causes (see LaMotta and Schiffer 2001). It is eminently possible to use the technology-transfer framework as a foundation for constructing myriad narratives. Needless to say, the precise form and content of any narrative will be tailored to the investigator's interests and causal preferences.[3] In the remainder of the chapter, it is shown how the technology-transfer framework is used to lay a behavioral foundation for constructing a historical narrative. In illustrating the foundational use of the framework, we draw on eighteenth-century electrical technologies.

The Case Study: Eighteenth-Century Electrical Technology

An ideal case study for present purposes is eighteenth-century electrical technology, because an extensive – and well-documented – process of differentiation took place in a relatively manageable timeframe, circa 1740–1810. Differentiation continued long after 1810, but this date is a convenient cutoff for the case study owing to the fundamentally new electrical technologies that arose in the early nineteenth century (see Schiffer 2001a).

Most people are surprised to learn that any electrical technology existed in the eighteenth century. Indeed, even some historians of electricity dispense quickly with eighteenth-century "static" or "electrostatic" technology,[4] tracing the beginnings of so-called practical electrical technology to Volta's (1800) invention of the electrochemical battery (e.g., Atherton 1984; Meyer 1971). However, electrical technology had a significant presence in Enlightenment societies and even found uses in the activities of elite and nonelite people in many European polities as well as in American colonies.

The technology-transfer framework suggested the questions that guided the research into this diverse and fascinating technology: (1) In what community did electrical technology originate and in which activities did it take part? (2) To which communities was electrical technology transferred?[5] (3) What groups made up the recipient communities? (4) In which activities of recipient communities did electrical technology participate? (5) What new functional variants arose in recipient communities and which performance characteristics were weighted?[6] and (6)

[3] For examples of behaviorally grounded narratives, see the explanations for the rise of Japanese consumer electronics in the 1950s and 1960s (Schiffer 1991, 1992, chap. 6), the failure of electric automobiles to trickle down to middle-class consumers in the first decades of the 20th century (Schiffer 2000, Schiffer et al. 1994a), and the adoption patterns of electric lighthouses in the 19th century (Schiffer 2005b).

[4] The terms static and electrostatic came into use only in the 19th century (e,g., Mascart 1876).

[5] The delineation of communities in the case study is somewhat simplified and incomplete (for a slightly different version, see Schiffer et al. 2003)

[6] To avoid becoming bogged down in excessive detail, the focus is on the more general performance characteristics (on general vs. specific performance characteristics, see Schiffer 2001a).

Which replication modes were used by members of recipient communities to acquire examples of the redesigned electrical technology?

Because electrical technology, even of the eighteenth century, was a complicated system consisting of dozens of parts, I regard any addition, deletion, or redesign of one or more of the system's parts as an instance of modification (i.e., the creation of a new variant). For present purposes, *electrical technology* is defined as any part, artifact, or system employed in the collection, generation, storage, distribution, manipulation, use, or display of electricity, that is, any *thing* that interacts directly or indirectly with electricity.

Electrophysicists

Electrical technology originated in the experimental activities of a small group of natural philosophers, in the community referred to as "electrophysicists." Among eighteenth-century electrophysicists were college and university professors, Jesuits and other clerics, medical doctors, and learned lay people such as Benjamin Franklin (Heilbron 1979). The electrophysicists also included, during use activities, a group of assistants – often unmentioned and rarely named in reports – who helped to manipulate the technology. By noting the research activities of several early electrophysicists, William Gilbert (1544–1600), Otto von Guericke (1602–1686), and Francis Hauksbee (1666–1713), we can introduce some fundamental principles of electrical technology.

Gilbert, court physician to Queen Elizabeth I, found that when substances such as amber, sulfur, and glass are rubbed, they acquired the ability to attract, at a distance, tiny, lightweight particles (Gilbert 1958:74–97). Today, we regard attraction as one manifestation of an electrical charge, which arises when an object has a relative excess or deficiency of electrons or other charge carriers. Friction is simply one way to create charges on an object's surface.

Mayor of Magdeburg, in Prussia, and inventor of a vacuum pump, Otto von Guericke (1994:227–231) constructed an electrical device in the late seventeenth century. It was a sulfur ball about 15 cm in diameter set in a wooden frame, which could be rubbed by hand and then turned 180° to display the "attractive virtue." In addition, he described electrical repulsion as well as the ability of charge to be communicated along a thread.

Francis Hauksbee (1719), employed as a demonstrator for the Royal Society in London, built the first electrical machine. It consisted of a glass globe mounted on an axle and connected by belt to a large, wooden driving wheel. With an assistant rotating the driving wheel by means of a crank, Hauksbee generated a charge merely by placing his hand against the spinning glass globe. Hauksbee carried out many important experiments, demonstrating among other effects that electricity could produce light – both as sparks and discharges in gases.

In the period 1720–1750, many additional effects were discovered that led to the construction of new electrical technologies (see Priestley 1966). The familiar principle that opposite charges attract and like ones repel was formulated, and conductors and insulators were distinguished. Investigators also learned that electricity can be readily "communicated" long distances by good conductors, such as metals, so long as they are insulated from the ground by glass supports or by suspension from silk threads. In addition, electrophysicists discovered that the electric "atmosphere" (today called a "field") surrounding a charged object could induce a charge on a nearby object, whether conductor or insulator. Finally, it was shown that some insulators, such as glass, could acquire and store for long periods a considerable charge; these were the first capacitors. The most widely adopted capacitor was the Leyden jar, a glass container often coated on the inside and outside with tin or lead foil. When charged by an electrical machine, the Leyden jar, or several of them wired in parallel as a "battery," could accumulate electricity sufficient to kill animals. Resting on these scientific and technological foundations, a recognizable electrical system was created by electrophysicists during the 1740s.

The basic system of that time consisted of an electrical machine and accessories (see Hackmann 1978). The heart of the electrical machine itself was a glass globe or cylinder, often 15–50cm in the greatest dimension, mounted on an axle and rotated by a crank. Against the glass vessel was placed a leather "rubber," usually coated with a mercury amalgam, which excited a charge on the glass. The charge was collected with a brass comb and conveyed to the "prime conductor," usually a closed brass tube on which the charge could be briefly stored. Glass supports, often coated with wax, insulated the electrical components from the wooden framework and ground. The most important accessory was a Leyden jar or battery, for it could accumulate large charges for use in experiments. Another essential accessory was the discharger, which was an insulated, tonglike device that could direct a charge from the Leyden jar, battery, or prime conductor to the object being electrified.

Electrophysicists also invented many specialized – sometimes unique – accessories, which usually consisted of common materials, including wood, metal, and glass, fashioned into new configurations, their mix of performance characteristics weighted for particular experiments. Even specialized accessories were copied by other investigators, and some were also commercialized by instrument makers.

Throughout the late eighteenth century, electrophysicists and members of other electrical communities also developed electrometers for indicating the "intensity" of the charge used in experiments (e.g., Bennet 1789; Cavallo 1777; Lane 1767). Because units of electrical measurement were neither well defined nor standardized in the eighteenth century (Bordeau 1982), *relative* intensity was usually denoted on an arbitrary scale based on spark length or degree of repulsion of like charges; in today's terminology, these devices indicated *voltage*. The concept equivalent to what is now called "current" emerged in the late eighteenth century; known as the "quantity" of electricity, it was usually indicated by the size and number of Leyden jars connected in parallel, the length of a metal wire of given diameter that the discharge could melt, spark thickness, or the severity of the shock received by the investigator. (The galvanometer, used for measuring quantity of current by its magnetic effects, was not invented until the early nineteenth century.)

Not unlike today's physicists, some eighteenth-century electrophysicists believed that bigger and more powerful electrical machines would disclose new effects. In the latter part of the century, as investigators sought higher power for experiments, they built and adopted new machine designs. The most successful design had a large, rotating glass plate in place of the former globe or cylinder (Hackmann 1978, Chap. 7). The largest plate machine, a one-of-a-kind creation, held two plates each about 1.5 m in diameter and required two or sometimes four men to turn the crank (Hackmann 1971). Commissioned by, and installed in, Teyler's Museum in Haarlem, Holland – where it still resides – this machine represents an early instance of what is now called "big science." An enormous and expensive apparatus, it supported collaborative, international work.

Compared with globe and cylinder machines, the larger plate machines had a mix of performance characteristics that weighted high-power output over both affordability and ease of use. This particular weighting was not favored in all communities, which is why the earlier machines continued to be replicated and acquired. Nonetheless, high-power plate machines were the most striking new variants of electrical technology produced by electrophysicists during the latter part of the eighteenth century.

Most scientists of the eighteenth century had many interests, and some conducted cutting-edge research in what, even then, was recognized as different disciplines. Joseph Priestley, for example, was both chemist and physicist. People like Priestley served as a principal mode for transferring information about electrical technology to communities outside physics. Typically, such individuals initiated the experimentation phase that could lead to the establishment of a new community. Information about electrical technology also moved with ease across disciplines as well as international borders because such transfers followed lines of communication that had been established during the late seventeenth and early eighteenth centuries by scientific societies and academies that embraced all disciplines (McClellan 1985).

Instrument makers, especially in London, rapidly commercialized electrical technology, offering electrical machines, basic accessories like the Leyden jar, and dozens of specialized accessories (for catalogs containing electrical technology, see Adams 1789, end of book, pp. 9–10; Adams and Jones 1799, 5:9–10; Adams et al. 1746:258–259; Guyot 1770:173; Nairne 1793:73–76). They also manufactured one-of-a-kind, custom apparatus for electrophysicists among others. Commonly, instrument makers published papers and experiment books that showcased their wares (e.g. Adams 1792; Adams and Jones 1799; Cuthbertson 1807; Guyot 1770; Nairne 1773, 1774, 1793; Neale 1747; Watkins 1747), and these served as additional modes for transferring information about electrical technology to people outside the community of electrophysicists. Although only the wealthiest individuals and institutions could afford the most expensive electrical machines, moderately priced and homemade machines enabled people, inside and outside science, to experiment with electricity and to judge the suitability of the technology for their own activities.

Electrical Demonstrators

Among the buyers of electrical machines were demonstrators who delivered public lectures, mainly to middle- and upper-class audiences (Millburn 1976; Schaffer 1983; Sutton 1995). Electrical demonstrators were a subset of an existing demonstrator–lecturer community, most of whose members rapidly acquired electrical technology. Electrical demonstrators rented halls, attracted audiences through handbills and newspaper advertisements, and displayed the latest electrical wonders. In this community, college and university lecturers are included because their public activities and electrical technologies were virtually identical to those of the itinerant electrical demonstrators. From the standpoint of use activities, assistants and audience members should also be regarded as community members. Indeed, responses of the audiences to electrical displays strongly influenced the design of new variants.

Exhibiting mastery over arcane forces, electrical demonstrators favored accessories that could arouse "wows" from an audience. Thus, they adopted devices having performance characteristics weighted so as to display the more striking effects that could attract attention and engage interest. A favorite was the "thunder-house," a wooden or paper model of a house that held a small charge of gunpowder. Upon the touch of a Leyden jar's conductor to the roof, the gunpowder exploded loudly, causing the model to fly apart as if struck by lightning. Other demonstrations included setting fire to "spirits of wine" (alcohol fumes), creating eerie displays of light, exploding fruit and pieces of wood, ringing small brass bells, and spelling out words in metal foils emblazoned with sparks (see, e.g., Ferguson 1778; Guyot 1770; Langenbucher 1780; Sigaud de la Fond 1776; Webers 1781). Among the more intriguing accessories were Ferguson's electrically powered models of clock, orrery, gristmill, and water pump (Ferguson 1778, Plates 2 and 3), which hinted at the possibility of electrically driven machinery.

By the end of the century, instrument makers had commercialized many dozens of accessories whose performance characteristics enabled electrical demonstrators to display many eye- and ear-catching effects in the context of educational and recreational activities justified by Enlightenment ideology (see Sutton 1995). Offerings included thunder-houses, bells, a porcelain head with hair that stood on end when electrified, and Ferguson's machinery models (e.g., Adams and Jones 1799, 5: catalog at end, p. 9). Some instrument makers even sold sets of demonstration apparatus along with carrying cases, obviously designed to be portable (for an early example, see Adams et al. 1746:259).

Collectors of Electrical Technology

Another important community consisted of wealthy people who bought electrical technologies for use and display at home, activities that conspicuously advertised their love of learning, as appropriate for members of the Enlightenment elite. Many of the same people who collected fossils, ancient artifacts, and other curiosities also had cabinets of philosophical and mathematical instruments. Electrical instruments were especially sought by collectors, and the richest among them commissioned the instrument makers to build custom machines, some even rivaling the monster at Teyler's Museum (Hackmann 1978). King George III and the Earl of Bute were among those who possessed collections of electrical things. It is believed by Daumas (1989:142) that collectors were the largest market for electrical devices in the eighteenth century. In addition to the collectors themselves, this community also consisted of family members and guests who took part in discussions about, and displays of, the electrical artifacts.

For guidance in the use of their new toys, collectors could purchase electrical manuals (with theory and experiment recipes) written by electrophysicists, electrical demonstrators, and instrument makers; by the 1780s, some volumes detailed more than 200 experiments (e.g., Cuthbertson 1807; Langenbucher 1780).

Electrical machines made for the collector market had to have, above all, stunning visual performance even when not in use, such as beautifully turned and varnished wooden elements, highly polished brass conductors, and glistening glass insulators. The most expensive machines were handsome and otherworldly creations – unlike any other artifacts of the eighteenth century. During the Enlightenment, the social competence of many an elite person apparently depended on the possession of an electrical machine and accessories that he or she could use and talk about knowledgeably (Daumas 1989:136–142). Indeed, electrical machines made for the elite, which could impress even the most sophisticated visitors, were instruments of social and political power.

Electrotherapists

Electrical technology of the Enlightenment also touched the lives – and bodies – of ordinary people. On the basis of experiments carried out on animals, including humans, investigators had shown that electricity could accelerate heart rates, increase perspiration, kill animals, and force muscles to contract (e.g., Jallabert 1748). Not surprisingly, given these physiological effects, a new community arose, that of electrotherapists (only some of whom were medical doctors) who employed electricity to treat sundry ailments. In activities of use, this community also included assistants and patients. Copious case histories presented in electrotherapy books indicate that diverse patients – young and old, rich and poor, male and female – received medical treatments (e.g., Cavallo 1780, 1795, Vol. 3; Lowndes 1787; Mauduyt 1784; Nairne 1793).

Although some early practitioners oversold electrical therapy (e.g., Lovett 1756), in later decades it was not pushed as a panacea. Rather, electrical treatment was recommended for particular maladies in which experience had seemingly demonstrated its efficacy, especially for cases unresponsive to more conventional therapies (e.g., Wesley 1792). One ailment sometimes ameliorated by electrical treatment was paralysis, such as that caused by stroke or by disuse of a limb after a severe accident. Applied repeatedly to atrophied muscles, electrical treatments could in some cases bring back the mobility of paralyzed limbs (e.g., Jallabert 1748; Sans 1772). Hysterical blindness and sundry "nervous disorders" were said as well to yield sometimes to electrical treatment. Surprisingly, revival of the "apparently dead" by cardiac stimulation was attempted with occasional success (e.g., Curry 1792; Kite 1788).

Three relatively benign techniques of electrification were used in electromedicine. In the first, individuals were placed on insulated stools, chairs, or beds and connected to the output of an electrical machine. This painless technique of charging a person, known as the "electric bath," was often used to treat nervous conditions and could last hours, day after day. The second technique made use of pointed metal or wood conductors to "draw sparks" from, or apply charge to, unwell parts of the body. These treatments were usually accomplished quickly, in a matter of minutes, but daily repetition was sometimes necessary. In the third technique, a moderate shock was directed, sometimes through long, jointed conductors, to specific organs or tissues. Such treatments, which in the case of paralysis might involve use of a Leyden jar, could last an hour or more and require many sessions.

Not surprisingly, new kinds of electrical machines and accessories were designed by and for members of the electromedical community (e.g., Blunt 1797; Cuthbertson 1807; Nairne 1783). Those made specifically for electromedicine, usually low-power cylinder machines, sometimes had prime conductors with sockets to receive plug-in accessories and a built-in Leyden jar (e.g., Nairne 1793). In addition, medical machines were compact for portability, and some were sold with carrying cases. Beyond the ubiquitous insulating stool, commercialized electromedical accessories included an array of tools for touching and approaching the human body (e.g., Adams 1792).

Four general performance characteristics seem to have been weighted in the design of medical accessories: (1) ability to draw sparks from, or conduct charge to, a specific part of the body (eyes, teeth, arm, and leg muscles, etc.); (2) ease of mechanical manipulation by the operator; (3) ability of the operator to employ the accessory without being shocked; and (4) capability of establishing a secure connection to the source of charge.

Atmospheric Electricians

Once Franklin and others had demonstrated that the atmosphere contained electricity, even on cloudless days, and that this charge could be drawn down to earth with suitable technology, a community formed to investigate variation in "atmospheric electricity."

Atmospheric electricians developed and adopted a number of technologies for collecting and sensing atmospheric charges. Because it was quickly learned that atmospheric electricity, although usually feeble, did vary directly with the height of one's collecting apparatus, kites were commonly adopted as a collecting technology (Bertholon 1787:33–81). Some investigators made large kites of their own design, and others modified the ubiquitous children's toy; often they oiled the paper to make it moisture resistant (e.g., Cavallo 1795, 2:5–6), but kites of cloth were said to be more durable (Bertholon 1787:322–323). They also placed a pointed metal collector on one of the sticks and included a brass or copper wire in the string to conduct the charge. A second appropriate technology was the hot-air balloon. Not long after the Montgolfier brothers' first flights with hot-air balloons in 1783, Bertholon had used one for studying atmospheric electricity (Bertholon 1787:332–334).

Still other investigators solved the collecting problem with more earth-bound technologies. Beccaria, for example, constructed an antennalike apparatus, connecting a long iron wire (insulated at both ends) from a chimney to the top of a cherry tree. From this collector, situated on Garzegna Hill in Mondavi, he ran a second wire into his laboratory below (Beccaria 1776:422–423). With terrestrial collectors, the charge was usually smaller than that obtained by a kite or balloon, and so investigators invented more sensitive indicators of intensity, such as Bennet's (1789) gold leaf electrometer, which instrument makers quickly commercialized (e.g., Adams and Jones 1799, 5: catalog at end, p. 10).

In devices designed for collecting atmospheric charges, two general performance characteristics seem to have been weighted: (1) the ability to collect a charge without endangering property or the investigator and (2) the ability to conduct the collected charge to a laboratory or a Leyden jar for later use. Performance characteristics weighted in electrometers designed for studying atmospheric electricity were sensitivity to charges of low intensity and, sometimes, portability.

The Property Protectors

Franklin was the first to propose that a building could be protected if a lightning strike was conducted directly to ground (e.g., Franklin 1996:124). Following Franklin's advice, some owners and managers, particularly people with oversight of churches, civic structures, powder magazines, and elite homes, formed another community referred to as "property protectors." Members of this community adopted what were then termed "lightning conductors" (see Bertholon 1787:265–267; Landriani 1784:240–242, appendix). A lightning conductor was usually an iron rod attached to a building, slightly above its highest point, which was connected by wire or other continuous metal pieces to moist ground. Although a few instrument makers commercialized standard lightning conductors (e.g., Adams and Jones 1799, 5: catalog at end, p. 10), most were probably custom made.

A large literature grew up around the proper design, installation, and use of lightning conductors, including detailed analyses of structures (with and without protection) that had been struck by lightning (e.g., Bertholon 1787; Blunt 1797;

Cavallo 1795, 1:78–82; Landriani 1784; Toaldo 1779; Wilson 1773). The principal general performance characteristics weighted in the design of lightning conductors were mechanical stability and the ability to conduct a large charge to ground. In more expensive designs, corrosion resistance was also weighted.

Electrochemists

The group of chemists who experimented with electrical technology for analysis and synthesis, along with their assistants, make up the community of electrochemists. There had been only modest activity in this community until the 1770s, when powerful disk machines became commercially available. Even so, electrochemists sometimes had to obtain access to the largest one-of-a-kind machines available (e.g., Priestley 1779:285); not surprisingly, the enormous machine at Teyler's Museum saw service in many electrochemical experiments (Levere 1969).

Several scientists, who were both chemists and electricians, transferred electrical technology to chemistry, creating accessories suitable for experimenting on gases (e.g., Cavallo 1781; Priestley 1781). In particular, they devised glassware with internal wires or electrodes in which reactions with gases could be stimulated by an electric spark. Such accessories were widely copied and, along with electrical machines and large batteries, entered many chemistry laboratories. Unique accessories were also created. For example, Cavallo constructed a metal combustion chamber that could be used when large explosions were anticipated (Cavallo 1781, Plate 2, Fig. 19). To ignite the gas, he employed a screw-in, glass-insulated electrode: the first spark plug (Cavallo 1781, Plate 2, Fig. 20).

Electrochemists also applied electricity to various liquid and solid substances (Nicholson 1795). To facilitate experiments such as the fusion of metal oxides into glasses, melting metals, and electrolysis of water, investigators constructed numerous small containers and specimen holders that many other members of the community adopted.

The performance characteristics of electrochemical accessories were weighted toward ensuring adequate containment of substances to be electrified as well as enabling a secure electrical connection to a Leyden jar or battery. In addition, owing to the large and potentially dangerous charges employed, the investigator had to be able to use these accessories in a "hands-off" mode.

Discussion and Conclusion

From its beginnings in the laboratories of electrophysicists, electrical technology was transferred rapidly to many other communities. By the 1780s, this technology had been adopted not only by several scientific communities, such as atmospheric electricians and electrochemists, but also by communities outside science, including the property protectors, instrument collectors, electrotherapists, and electrical demonstrators. In the course of these intercommunity transfers, the technology became greatly differentiated, as people in recipient communities invented new functional variants whose performance characteristics were tailored to the situational factors of their activities. Many of these new designs were commercialized by instrument makers, who made them readily available to members of diverse communities.

The technology-transfer framework has obvious implications for studying technology in complex societies, like the case study, in which it is relatively easy to delineate behaviorally based communities. Anthropologists who study less complex societies might be wondering if the technology-transfer framework could be of use to them. Two points argue in favor of its general applicability.

First, technological differentiation occurs in societies at all levels of complexity. Thus, the technology-transfer framework applies, by definition, to any case of technological differentiation, regardless of time, place, or nature of the society or technology. Second, because *community* is defined in the most general terms, investigators can delineate communities in any society. In less complex societies, for example, the investigator can use identifiers such as sex, gender, age, wealth, developmental stage of household, and place of residence to designate communities provisionally. After further study, it should be possible to define these communities behaviorally; after all, gender- and age-based roles tend to have some corresponding activities and technologies. Even part-time occupational specialization can furnish a behavioral basis for defining communities.

Many of the communities recognizable in simpler societies are apt to have greatly overlapping memberships. Thus, the community of farmers might share most of its members with the woodworking community, and the everyday cooking community might overlap largely with the community of potters. This is not a problem in the present framework because community memberships can, by definition, overlap to any extent. Even in the case study, as the references show, there were considerable membership overlaps among eighteenth-century electrical communities. In principle, then, the technology-transfer framework is flexible enough to handle technological differentiation in the most socially homogeneous societies.

How does one apply this framework to new cases? In the first step, the investigator specifies the technology of interest. Once the technology is defined, it is necessary to document its design variants, mapping out, at least coarsely, time–space coordinates of the functional variation. Next, the investigator employs diverse lines of evidence to delineate provisional communities and user groups associated with the adopted variants and the activities in which the latter took part. Although one can attempt to model the specific information transfers that occurred among communities, this level of detail is not essential. The next step is to infer the weighting of the performance characteristics of the redesigned variants in relation to situational factors and relevant user groups. For example, a community whose food-serving bowls are seen by visitors would be expected to adopt different designs than a community whose food-serving bowls are never in view. The behavioral theory of design (Chap. 1; Schiffer and Skibo 1997; Skibo and Schiffer 2001) plays an important role in constructing proximate explanations of redesign in the context of the adopting community's activities. At last, the investigator constructs on this behavioral foundation a richly textured historical narrative.

Whether one is interested in the proliferation of chipped-stone technologies during the Middle and Upper Paleolithic, the differentiation of Pueblo ritual technologies, or the expansion of industrial clothing technologies in the twentieth century, the technology-transfer framework provides a potentially useful tool for anthropologists studying large-scale patterns of technological change and variability. Clearly, this tool enables investigators to confront and handle, in very specific terms, the materiality of human activities. By building on this behavioral foundation, we can craft historical narratives of technological differentiation that are more than just-so stories.

References

Ackerly, N. W., J. B. Howard, and R. H. McGuire 1987 *La Ciudad Canals: A Study of Hohokam Irrigation Systems at the Community Level*. Anthropological Field Studies, No. 17, La Ciudad Monograph Series, Vol. 2. Arizona State University, Office of Cultural Resource Management, Tempe, Arizona.

Adams, G. 1789 *An Essay on Vision*. London.

Adams, G. 1792 *An Essay on Electricity, Explaining the Principles of That Useful Science; and Describing the Instruments, Contrived Either to Illustrate the Theory, or Render the Practice Entertaining*. 4th edition. London.

Adams, J. 1994 The development of prehistoric grinding technology in the Point of Pines area. Ph. D. dissertation, University of Arizona. University Microfilms, Ann Arbor.

Adams, J. 1999 Refocusing the role of food-grinding tools as correlates for subsistence strategies in the U.S. Southwest. *American Antiquity* 64: 475–498.

Adams, G., and W. Jones 1799 *Lectures on Natural and Experimental Philosophy, Considered in Its Present State of Improvement*. 2nd edition. 5 Vols. London.

Adams, G., L. Joblot, and A. Trembley 1746 *Micrographia Illustrata; or, the Knowledge of the Microscope Explained*. London.

Aikens, C. M. 1995 First in the world: Jomon potter of early Japan. In *The Emergence of Pottery: Technology and Innovation in Ancient Societies*, edited by W. K. Barnett and J. W. Hoopes, pp. 11–21. Smithsonian Institution Press, Washington, D. C.

Anonymous. 1862 Electric fire telegraphs on the continent. *The Engineer* 14: 185.

Appadurai, A. (editor) 1986 *The Social Life of Things*. Cambridge University Press, Cambridge.

Arnold, D. E. 1985 *Ceramic Theory and Cultural Process*. Cambridge University Press, Cambridge.

Arnold, D. E. 1993 *Ecology of Ceramic Production in an Andean Community*. Cambridge University Press, Cambridge.

Arnold, P. J. III 1991 *Domestic Ceramic Production and Spatial Organization: A Mexican Case Study in Ethnoarchaeology*. Cambridge University Press, Cambridge.

Arnold, P. J. 1999a On typologies, selection, and ethnoarchaeology in ceramic production studies. In *Material Meanings: Critical Approaches to the Interpretation of Material Culture*, edited by E. S. Chilton, pp. 103–117. University of Utah Press, Salt Lake City, Utah.

Arnold, P. J. 1999b Tecomates, Residential Mobility, and Early Formative occupation in coastal lowland Mesoamerica. In *Pottery and People*, edited by J. M. Skibo and G. M. Feinman, pp. 159–170. University of Utah Press, Salt Lake City, Utah.

Aronson, M., J. M. Skibo, and M. T. Stark 1994 Production and use technologies in Kalinga pottery. In *Kalinga Ethnoarchaeology: Expanding Archaeological Method and Theory*, edited by W. A. Longacre and J. M. Skibo, pp. 83–112. Smithsonian Institution Press, Washington, D.C.

Arthur, J. W. 2002 Pottery use-alteration as an indicator of socioeconomic status: An ethnoarchaeological study of the Gamo of Ethiopia. *Journal of Archaeological Method and Theory* 9: 331–355.

Arthur, J. W. 2003 Brewing beer: Status, wealth, and ceramic use-alteration among the Gamo of South western Ethiopia. *World Archaeology* 34: 516–528.

Arthur, J. W. 2007 *Living with Pottery*. University of Utah Press, Salt Lake City, Utah.

Atherton, W. A. 1984 *From Compass to Computer: A History of Electrical and Electronics Engineering*. San Francisco Press, San Francisco.

Bamforth, D. B. 2002 Evidence and metaphor in evolutionary archaeology. *American Antiquity* 67: 435–452.

Barlow, P. 1825 On the laws of electro-magnetic action, as depending on the length and dimensions of the conducting wire, and on the question, whether electrical phenomena are due to the transmission of a single or compound fluid. *Edinburgh Philosophical Journal* 12: 105–114.

Barnett, W. K., and J. W. Hoopes (editors) 1995 *The Emergence of Pottery: Technological Innovation in Ancient Societies*. Smithsonian Institution Press, Washington, D. C.

Basalla, G. 1988 *The Evolution of Technology*. Cambridge University Press, Cambridge.

Beauchamp, K. 2001 *History of Telegraphy*. The Institution of Electrical Engineers, London.

Beccaria, G. 1776 *A Treatise upon Artificial Electricity*. London.

Beck, M. E., and M. E. Hill 2004 Rubbish, relatives, and residence: The family use of middens. *Journal of Archaeological Method and Theory* 11: 297–333.

Beck, M. E., and M. E. Hill 2007 Midden ceramics and their sources in Kalinga. In *Archaeological Anthropology: Perspectives on Method and Theory*, edited by J. M. Skibo, M. W. Graves, and M. T. Stark, pp. 111–137. University of Arizona Press, Tucson.

Beck, M. E., J. M. Skibo, D. Hally, and P. Yang 2002 Use-alteration of abraded sherds from the Late Archaic. *Journal of Archaeological Science* 29: 1–15.

Bédoucha, G. 2002 The watch and the waterclock: Technological choices/social choices. In *Technological Choices: Transformation in Material Cultures since the Neolithic*, edited by P. Lemonnier, pp. 77–107. Routledge, London.

Beld, S. 2001 The Cater site (20MD36): Description of excavations and recovered material. In *The Cater Site: Archaeology, History, Artifacts, and Activities at This Early 19th Century Midland County Site*, edited by D. J. Fruip, pp. 14–41. Chippewa Nature Center, Midland, Michigan.

Bennet, A. 1789 *New Experiments on Electricity*. Derby.

Bertholon, P. 1787 *De l'Électricité des Météores. 2 Vols. Bernuset*, Lyon.

Bijker, W. E. 1995 *Of Bicycles, Bakelites, and Bulbs: Toward a Theory of Sociotechnical Change*. MIT, Cambridge, Massachusetts.

Binford, L. R. 1967 Smudge pits and hide smoking: The use of analogy in archaeological reasoning. *American Antiquity* 31: 203–210.

Binford, L. R. 1968 Archeological perspectives. In *New Perspectives in Archeology*, edited by S. R. Binford and L. R. Binford, pp. 5–32. Aldine, Chicago.

Binford, L. R. 1972 *An Archaeological Perspective*. Seminar Press, New York.

Binford, L. R., J. Schoenwetter, and M. L. Fowler 1964 *Archaeological Investigations in the Carlyle Reservoir*. Archaeological Salvage Report No. 17, pp. 1–117. Southern Illinois University Museum, Carbondale.

Bishop, C. H. 1974 *The Northern Ojibwa and the Fur Trade: A Historical and Ecological Study*. Holt, Rinehart, and Winston, Toronto.

Bleed, P. 2001a Trees or chains, links or branches: Conceptual alternatives for the consideration of stone tool production and other sequential activities. *Journal of Archaeological Method and Theory* 8: 101–127.

Bleed, P. 2001b Artifice constrained: What determines technological choice? In *Anthropological Perspectives on Technology*, edited by M. B. Schiffer, pp. 151–162. University of New Mexico Press, Albuquerque.

Blinman, E., and C. D. Wison 1994 Additional ceramic analyses. In *Across the Colorado Plateau: Anthropological Studies for the Trans-western Pipeline Expansion Project*, Vol. 14. Office of Contract Archaeology and Maxwell Museum of Anthropology, University of New Mexico, Albuquerque.

Blunt, T. 1797 *Description and Use of Blunt's Medical Electric Machine, and a New Method of Applying Metallic Conductors to Buildings, &C. for their Preservation from Lightning*. London.

Bollong, C. A., J. C. Vogel, L. Jacobson, W. A. Van der Westhuizen, and C. G. Sampson 1993 Direct dating and identity of fibre temper in pre-contact Bushman (Basarwa) pottery. *Journal of Archaeological Science* 20: 41–55.

Boone, J. L., and E. A. Smith 1998 Is it evolution yet? A critique of evolutionary archaeology. *Current Anthropology* 39: 141–173.

Bordeau, S. P. 1982 *Volts to Hertz. The Rise of Electricity from the Compass to the Radio through the Works of Sixteen Great Men of Science Whose Names Are Used in Measuring Electricity and Magnetism*. Burgess, Minneapolis.

Bourdieu, P. 1977 *Outline of a Theory of Practice*. Cambridge University Press, Cambridge.

Bourdieu, P. 1990 *The Logic of Practice*. Polity Press, Cambridge.

Braun, D. P. 1983 Pots as tools. In *Archaeological Hammers and Theories*, edited by A. Keene and J. Moore, pp. 107–134. Academic, New York.

Breternitz, D. A. 1982 The four corners Anasazi ceramic tradition. In "Southwestern ceramics: A comparative review" edited by A. H. Schroeder, pp. 129–147. The Arizona Archaeologist 15. School of American Research, Santa Fe.

Brotherton, R. A. 1944 Discovery of iron ore: Negaunee centennial (1844–1944). *Michigan History* 28: 199–213.

Broughton, J. M., and J. F. O'Connell 1999 On evolutionary ecology, selectionist archaeology, and behavioral archaeology. *American Antiquity* 64: 153–165.

Brown, J. A. 1989 The beginnings of pottery as an economic process. In *What's New? A Closer Look at the Process of Innovation*, edited by S. E. van der Leeuw and R. Torrence, pp. 203–224. Unwin Hyman, London.

Buffalohead, P. K. 1983 Farmers, warriors, traders: A fresh look at Ojibway women. *Minnesota History* 48: 236–244.

Burton, J. F. 1991 *The Archaeology of Sivu'ovi: The Archaic to Basketmaker Transition at Petrified Forest National Park*. Publications in Anthropology No. 55. Western Archeological and Conservation Center, National Park Service, Tucson, Arizona.

Castle, B. H. 1987 *The Grand Island Story*. The John M. Longyear Research Library, Marquette, Michigan.

Catlin, G. 1985 *Letters and Notes on the Manners, Customs, and Conditions of the North American Indians: Written During Eight Years' Travel (1832–1839) Amongst The Wildest Tribes of Indians in North America*, Vol. 1. Dover Publications, New York.

Cavallo, T. 1777 New electrical experiments and observations; with an improvement of Mr. Canton's electrometer. *Philosophical Transactions of the Royal Society* 67: 388–400.

Cavallo, T. 1780 *An Essay on the Theory and Practice of Medical Electricity*. London.

Cavallo, T. 1781 *A Treatise on the Nature and Properties of Air, and Other Permanently Elastic Fluids*. London.

Cavallo, T. 1795 *A Complete Treatise on Electricity, in Theory and Practice; with Original Experiments*. 4th edition. 3 Vols. London.

Chandler, A. D., Jr. 1977 *The Visible Hand: The Managerial Revolution in American Business*. Belknap Press, Cambridge, Massachusetts.

Channing, W. F. 1855 The American Fire-Alarm Telegraph. *Smithsonian Institution, Ninth Annual Report*, pp. 147–155. U.S. Government Printing Office, Washington, D. C.

Chapman, J., and B. Gaydarska 2007 *Parts and Wholes: Fragmentation in Prehistoric Context*. Oxbow Books, Oxford.

Childe, V. G. 1951 *Man Makes Himself*. New American Library of World Literature, London.

Clark, J. E., and D. Gosser 1995 Reinventing Mesoamerica's first pottery. In *The Emergence of Pottery: Technology and Innovation in Ancient Societies*, edited by W. K. Barnett and J. W. Hoopes, pp. 209–221. Smithsonian Institution Press, Washington, D. C.

Cobb, C. R., and A. King 2005 Re-inventing Mississippian tradition at Etoway, Georgia. *Journal of Archaeological Method and Theory* 12: 167–192.

Conkey, M. W. 2007 Questioning theory: Is there a gender of theory in archaeology? *Journal of Archaeological Method and Theory* 14: 285–310.

Cooper, C. C. 1991 *Shaping Invention: Thomas Blanchard's Machinery and Patent Management in Nineteenth-Century America*. Columbia University Press, New York.

Cordell, L. 1997 *Archaeology of the Southwest*. 2nd edition. Academic Press, San Diego.

Costin, C. L. 1991 Craft specialization: Issues in defining, documenting, and explaining the organization of production. *Archaeological Method and Theory* 3: 1–56.

Coulam, N. J., and A. R. Schroedl 1996 Early Archaic clay figurines from Cowboy and Walters Caves in southwestern Utah. *Kiva* 61: 401–412.

Cresswell, R. 1976 Avant-propos. *Techniques et Culture* 1: 5–6

Cresswell, R. 2002 Of mills and waterwheels: The hidden parameters of technological choice. In *Technological Choices: Transformation in Material Cultures since the Neolithic*, edited by P. Lemonnier, pp. 181–213. Routledge, London.

Crown, P. L., 1994 *Ceramics and Ideology: Salado Polychrome Pottery*. University of New Mexico Press, Albuquerque.

Crown, P. L., and W. H. Wills 1995 Economic intensification and the origins of ceramic containers in the American Southwest. In *The Emergence of Pottery: Technology and Innovation in Ancient Societies*, edited by W. K. Barnett and J. W. Hoopes, pp. 241–254. Smithsonian Institution Press, Washington, D. C.

Curot, M. 1911 Curot's Journal, 1803. In *Collections of the State Historical Society of Wisconsin, Vol. XX: The Fur Trade in Wisconsin–1812–1825, and a Wisconsin Trader's Journal 1803–04*, edited by R. G. Thwaites. State Historical Society of Wisconsin, Milwaukee.

Curry, J. 1792 *Popular Observations on Apparent Death from Drowning, Suffocation, & C. with an Account of the Means to Be Employed for Recovery*. London.

Cuthbertson, J. 1807 *Practical Electricity, and Galvanism, Containing a Series of Experiments Calculated for the Use of Those Who Are Desirous of Becoming Acquainted with That Branch of Science*. London.

Daumas, M. 1989 *Scientific Instruments of the 17th and 18th Centuries and Their Makers*. Portman Press, London.

Dart, A. 1989 *Prehistoric Irrigation in Arizona: a Context for Canals and Related Cultural Resources*. Arizona State Historic Preservation Office, Phoenix.

David, N., and C. Kramer 2000 *Ethnoarchaeology in Action*. Cambridge University Press, Cambridge.

Deal, M. 1998 *Pottery Ethnoarchaeology in the Central Mayan Highlands*. University of Utah Press, Salt Lake City, Utah.

Densmore, F. 1979 *Chippewa Customs*. Minnesota Historical Society Press, St. Paul.

Dethlefsen, E., and J. Deetz 1966 Death's heads, cherubs and willow trees: Experimental archaeology in colonial cemeteries. *American Antiquity* 31: 502–510.

Dibner, B. 1959 The Atlantic Cable. *Burndy Library, Publication* No. 16. Norwalk, Connecticut.

Dibner, B. 1961 Oersted and the Discovery of Electromagnetism. *Burndy Library, Publication* No. 18. Norwalk, Connecticut.

Dietler, M., and I. Herbich 1996 *Habitus*, techniques, style: An integrated approach to the social understanding of material culture and boundaries. In *The Archaeology of Social Boundaries*, edited by M. T. Stark, pp. 232–263. Smithsonian Institution Press, Washington, D. C.

Di Peso, C. C. 1974 *Casas Grandes: A Fallen Trade Center of the Gran Chichimeca*, Vols. 1–3. Amerind Foundation, Dragoon, Arizona, Northland Press, Flagstaff.

Di Peso, C. C., J. B. Rinaldo, and G. J. Fenner 1974 *Casas Grandes: A Fallen Trade Center of the Gran Chicimeca*, Vols. 4–8. Amerind Foundation, Dragoon, Arizona, Northland Press, Flagstaff.

Dittert, A. E., Jr., J. J. Hester, and F. W. Eddy 1961 *An Archaeological Survey of the Navajo Reservoir District, Northwestern New Mexico*. Monographs of the School of American Research, No. 23. Santa Fe, New Mexico.

Dobres, M.-A. 1995 Gender and prehistoric technology: On the social agency of technical strategies. *World Archaeology* 27: 25–49.

Dobres, M.-A. 2000 *Technology and Social Agency: Outlining a Practice Framework for Archaeology*. Blackwell Publishers, Oxford.

Dobres, M.-A. 2001 Meaning in the making: agency and the social embodiment of technology and art. In *Anthropological Perspectives on Technology*, edited by M. B. Schiffer, pp. 47–76. University of New Mexico Press, Albuquerque.

Dobres, M.-A., and C. R. Hoffman 1994 Social agency and the dynamics of prehistoric technology. *Journal of Archaeological Method and Theory* 1: 211–258.

Dobres, M.-A., and C. R. Hoffman (editors) 1999 *The Social Dynamics of Technology: Practice, Politics, and World Views*. Smithsonian Institution Press, Washington, D. C.

Dobres, M.-A., and J. E. Robb 2000 Agency in archaeology: Paradigm or platitude? In *Agency in Archaeology*, edited by M.-A. Dobres and J. E. Robb, pp. 3–17. Routledge, London.

Dobres, M.-A., and J. E. Robb 2005 "Doing" agency: Introductory remarks on methodology. *Journal of Archaeological Method and Theory* 12: 159–166.

Dornan, J. L. 2002 Agency and archaeology: Past, present, and future directions. *Journal of Archaeological Method and Theory* 9: 303–329.

Douglas, J. 1995 Autonomy and regional systems in the late prehistoric southern Southwest. *American Antiquity* 60: 240–257.

Douglas, M., and B. Isherwood 1979 *The World of Goods: Towards an Anthropology of Consumption*. Basic Books, New York.

Drake, E., and M. Drake 2007 Logging, labor, and landscapes: Transformations in labor relations and the spatial organization of the late 19th and early 20th century logging camps in the Upper Great Lakes. Paper presented at the Annual Meeting of the Society for Historical Archaeology.

Drake, E., J. G. Franzen, and M. Drake 2006 Working at home and living at work: Negotiating tensions of work, leisure, and place in the late 19th and early 20th century logging camps. Paper presented at the Annual Meeting of the Society for American Archaeology, San Juan.

Drake, M. n.d. The Social and Historical Context of Finnish Savusaunas on 20th Century Logging Camps in the Upper Peninsula of Michigan: A Case study in Historical Archaeology. M.A. Thesis, Department of Anthropology, Western Michigan University, Kalamazoo.

Dunham, S. B., and J. B. Anderton 1999 Late archaic radiocarbon dates from the Popper site (FS 09-10-03-825/20AR350): A multicomponent site on Grand Island, Michigan. *Michigan Archaeologist* 45: 1–22.

Dunham, S. B., and M. C. Branstner 1995 *1994 Phase II Cultural Resource Evaluations: Hiawatha National Forest*. USDA-Forest Service, Hiawatha National Forest, Escanaba, Michigan. Great Lakes Research Associates, Williamston, Michigan (GLRA Report No. 94-05).

Dunnell, R.C. 1978 Style and function: A fundamental dichotomy. *American Antiquity* 43:192–202.

Dunnell, R. C. 1980 Evolutionary theory and archaeology. *Advances in Archaeological Method and Theory* 3: 35–99.

Dunnell, R. C. 1982 Science, social science, and common sense: The agonizing dilemma of modern archaeology. *Journal of Anthropological Research* 38: 1–25.

Dunnell, R. C. 1989 Aspects of the application of evolutionary theory in archaeology. In *Archaeological Thought in America*, edited by C. C. Lamberg-Karlovsky, pp. 35–49. Cambridge University Press, New York.

Dunnell, R. C., and J. K. Feathers 1991 Late Woodland manifestations of the Malden Plain, Southwest Missouri. In *Stability, Transformation, and Variation: The Late Woodland Southeast*, edited by M. S. Nassaney and C. R. Cobb, pp. 21–45. Plenum, New York

Eddy, F. W. 1966 *Prehistory in the Navajo Reservoir District, Northwestern New Mexico*. Museum of New Mexico Papers in Anthropology No. 15. Santa Fe, New Mexico.

Eighmy, J. L., and J. B. Howard 1991 Direct dating of prehistoric canal sediments using archaeomagnetism. *American Antiquity* 56: 88–102.

Fahie, J. J. 1884 *A History of Electric Telegraphy, to the Year 1837*. E. & F.N. Spon, London.

Feathers, J. K. 2006 Explaining shell-tempered pottery in prehistoric eastern North America. *Journal of Archaeological Method and Theory* 13: 89–133.

Ferguson, J. 1778 *An Introduction to Electricity in Six Sections*. 3rd edition. London.

Fewkes, J. W. 1892 The ceremonial circuit among the village Indians of Northwestern Arizona. *Journal of American Folk-Lore* 5(16):33–42.

Finn, B. S. 1973 *Submarine Telegraphy: The Grand Victorian Technology*. Science Museum, London.

Fish, P. R., and S. K. Fish 1999 Reflections of the Casas Grandes regional system from the northwestern periphery. In *The Casas Grandes World*, edited by C. F. Schaafsma and C. L. Riley, pp. 27–42. University of Utah Press, Salt Lake City, Utah.

Fitzhugh, B. 2001 Risk and invention in human technological evolution. *Journal of Anthropological Archaeology* 20: 125–167.

Fowler, A. P. 1991 Brown ware and red ware pottery: An Anasazi ceramic tradition. *Kiva* 56: 123–144.

Franklin, B. 1996 (1769) *Experiments and Observations on Electricity, Made at Philadelphia in America*. Classics of Science Library, New York.

Franzen, J. G. 1992 Northern Michigan logging camps: Material culture and worker adaptation on the industrial frontier. *Historical Archaeology* 26: 74–98.

Franzen, J. G. 1995 Comfort for man or beast: Alcohol and medicine use in Northern Michigan logging camps, ca. 1880–1940. *The Wisconsin Archeologist*. 76: 294–337.

Galison, P. 2003 *Einstein's Clocks; Poincaré's Maps: Empires of Time*. Norton, New York.

Gebauer, E. J. 1995 Pottery production and the introduction of agriculture in southern Scandinavia. In *The Emergence of Pottery: Technology and Innovation in Ancient Societies*, edited by W. K. Barnett and J. W. Hoopes, pp. 99–112. Smithsonian Institution Press, Washington, D. C.

Giddens, A. 1979 *Central Problems in Social Theory: Action, Structure, and Contradiction in Social Analysis*. University of California Press, Berkeley.

Giddens, A. 1984 *The Constitution of Society: Outline of a Theory of Structuration*. University of California Press, Berkeley.

Gilbert, W. 1958 (1600) *De Magnete*. (Reprint of English translation by P. Fleury Mottelay, first published in 1893.) Dover, Mineola, New York.

Gillespie, S. D. 1991 Ballgames and boundaries. In *The Mesoamerican Ballgame*, edited by V. L. Scarborough and D. R. Wilcox, pp. 317–346. University of Arizona Press, Tucson.

Gilman, C. R. 1836 *Life on the Lakes Being Tales and Sketches Collected During a Trip to the Pictured Rocks of Lake Superior*. George Dearborn, New York.

Gilman, P. A. 1987 Architecture as artifact: Pit structures and pueblos in the American southwest. *American Antiquity* 52: 538–564.

Gilman, R. R. 1974 The fur trade in the upper Mississippi valley, 1630–1850. *Wisconsin Magazine of History*, 58: 3–18.

Gordon, R. B., and D. Killick 1993 Adaptation of technology to culture and environment: Bloomery iron smelting in America and Africa. *Technology and Culture* 34: 243–270.

Gosselain, O. P. 1998 Social and technical identity in a clay crystal ball. In *The Archaeology of Social Boundaries*, edited by M. T. Stark, pp. 78–106. Smithsonian Institution Press, Washington, D. C.

Gould, R. A. 2001 From sail to steam at sea in the late nineteenth century. In *Anthropological Perspectives on Technology*, edited by M. B. Schiffer, pp. 193–213. University of New Mexico Press, Albuquerque.

Graham, L. R. 1995 *A Face in the Rock: The Tale of the Grand Island Chippewa*. Island Press, Washington, D. C.

Graves, M. W., and T. N. Ladefoged 1995 The evolutionary significance of ceremonial architecture in Polynesia. In *Evolutionary Archaeology: Methodological Issues*, edited by P. A. Teltser, pp. 149–174. University of Arizona Press, Tucson.

Great Britain Patent Office 1859 *Patents for Inventions. Abridgments of Specifications Relating to Electricity and Magnetism, their Generation and Applications*. Great Britain Patent Office, London.

Great Britain Patent Office 1874 *Patents for Inventions. Abridgments of Specifications Relating to Electricity and Magnetism, their Generation and Application. Part II—A.D. 1858–1866*. 2nd edition. Great Britain Patent Office, London.

Great Britain Patent Office 1882 *Patents for Inventions. Abridgments of Specifications Relating to Electricity and Magnetism. Division II. Conducting and Insulating. Part II—A.D. 1867–1876*. Great Britain Patent Office, London.

Green, R. C. 1979 Lapita Horizon. In *The Prehistory of Polynesia*, edited by J. D. Jennings, pp. 27–60. Harvard University Press, Cambridge, Massachusetts.

Guericke, Otto von 1994 (1672) The New (So-Called) Magdeburg Experiments of Otto von Guericke. M. G. F. Ames, trans. Kluwer, Dordrecht.

Gumerman, G. J. (editor) 1991 *Exploring the Hohokam: Prehistoric Desert Peoples of the American Southwest*. University of New Mexico Press, Albuquerque.

Guyot, M. 1770 *Nouvelle Recreations Physiques et Mathématiques*, Vol. 2. Nouvelle Edition. Paris.

Hackmann, W. D. 1971 Electrical researches. In *Martinus van Marum: Life and Work*, edited by R. J. Forbes, Vol. 3, pp. 329–378. Tjeenk Willink and Zoon, Haarlem, Holland.

Hackmann, W. D. 1978 *Electricity from Glass: The History of the Frictional Electrical Machine 1600–1850*. Sijthoff and Noordhoff, The Netherlands.

Hally, D. J. 1983 Use alteration of pottery vessel surfaces: An important source of evidence for the identification of vessel function. *North American Archaeologist* 4: 3–26.

Hardwick, N. 2008 The Underhill Camp: Exploring the Intrasite Spatial Arrangement of an Early Twentieth-Century Logging Site Using a Performance-Based Life History Approach to Link Archaeological Patterns to People and Practices. M.A. Thesis, Illinois State University, Normal, Illinois.

Hart, J. P., and J. E. Terrell 2002 *A Handbook of Concepts in Modern Evolutionary Archaeology*. Greenwood Press, Westport, Connecticut.

Hauksbee, F. 1719 *Physico-Mechanical Experiments on Various Subjects. Containing an Account of Several Surprizing Phenomena Touching Light and Electricity, Producible on the Attrition of Bodies*. London.

Haury, E. W. 1976 *Hohokam Desert Farmers and Craftsmen: Excavations at Snaketown, 1964–1965*. University of Arizona Press, Tucson.

Haury, E. W. 1985 An early pit house village of the Mogollon culture Forestdale valley, Arizona. In *Mogollon Culture in the Forestdale Valley, East-Central Arizona*, edited by E. W. Haury, pp. 285–371. University of Arizona Press, Tucson.

Hayden, B. 1993 *Archaeology: The Science of Once and Future Things*. W. H. Freeman, New York.

Hayden, B. 1995 The emergence of prestige technologies and pottery. In *The Emergence of Pottery: Technology and Innovation in Ancient Societies*, edited by W. K. Barnett and J. W. Hoopes, pp. 257–264. Smithsonian Institution Press, Washington, D. C.

Hayden, B. 1998 Practical and prestige technologies: The evolution of material systems. *Journal of Archaeological Method and Theory* 5: 1–55.

Hays, K. 1992 Anasazi ceramics as text and tool: Toward a theory of ceramic design "messaging." Ph. D. dissertation, Department of Anthropology, University of Arizona, Tucson.

Headrick, D. R. 1981 *The Tools of Empire: Technology and European Imperialism in the Nineteenth Century*. Oxford University Press, Oxford.

Hegmon, M. 2003 Setting theoretical egos aside: Issues and theory in North American archaeology. *American Antiquity* 68: 213–244.

Heidke, J. M., E. Miksa, and M. K. Wiley 1997 Ceramic artifacts. In *Archaeological Investigation of Early Village Sites in the Middle Santa Cruz Valley*, edited by J. B. Mabry. Anthropological Papers No. 19. Center for Desert Archaeology, Tucson, Arizona.

Heilbron, J. L. 1979 *Electricity in the 17th and 18th Centuries: A Study in Early Modern Physics*. University of California Press, Berkeley.

Henry, J. 1831 On the application of the principle of the galvanic multiplier to electro-magnetic apparatus, and also to the developement [sic] of great magnetic power in soft iron. *The American Journal of Science and Arts* 19: 400–408.

Hilger, M. I. 1992 *Chippewa Child Life and its Cultural Background*. Minnesota Historical Society Press, St. Paul.

Hill, J. N. 1970 *Broken K Pueblo: Prehistoric Social Organization in the American Southwest*. Anthropological Papers No. 18. University of Arizona Press, Tucson.

Hiscox, G. D. 1900 *Horseless Vehicles, Automobiles, Motorcycles Operated by Steam, Hydro-Carbon, Electric, and Pneumatic Motors*. Munn, New York.

Hodder, I. 1985 Postprocessual archaeology. In *Advances in Archaeological Method and Theory* 8:1–26.

Hodder, I., and S. Hutson 2003 *Reading the Past: Current Approaches to Interpretation in Archaeology*. Cambridge University Press, Cambridge.

Holmberg, J. W. 1969 *Nomads of the Long Bow: The Siriono of Eastern Bolivia*. The American Museum of Science Books, New York.

Huckell, B. B. 1990 Late preceramic farmer-foragers in Southeastern Arizona: A cultural and ecological consideration of the spread of agriculture into the arid Southwestern United States. Ph. D. dissertation, Arid Lands Resource Sciences, University of Arizona, Tucson.

Huckleberry, G. 1999 Assessing Hohokam canal stability through stratigraphy. *Journal of Field Archaeology* 26: 1–18.

Hughes, S. 1998 Getting to the point: Evolutionary change in prehistoric weaponry. *Journal of Archaeological Method and Theory* 5: 345–403.

Hughes, T. 1983 *Networks of Power: Electrification in Western Society 1980–1930*. Johns Hopkins University Press, Baltimore.

Hughes, T. 1990 From deterministic dynamos to seamless-web systems. In *Engineering as a Social Enterprise*, edited by H. Sladovich, pp. 7–25. National Academy, Washington, D. C.

Hurt, T. D., and G. F. M. Rakita (editors) 2001 *Style and Function: Conceptual Issues in Evolutionary Archaeology*. Bergin and Garvey, Westport, Connecticut.

Hutchins, E. 1995 *Cognition in the Wild*. MIT, Cambridge, Massachusetts.

Israel, P. B. 1989 *From the Machine Shop to the Industrial Laboratory: Telegraphy and the Changing Context of American Invention, 1830–1920*. University Microfilms, Ann Arbor, Michigan.

Israel, P. B. 1998 *Edison: A Life of Invention*. Wiley, New York.

Jallabert, J. 1748 *Experiénces sur l'Électricité, avec Quelques Conjectures sur la cause de ses Effets*. Geneva.

Jarmon, N. 1998 Material of culture, fabric of identity. In *Material Cultures: Why Some Things Matter*, edited by D. Miller, pp. 121–146. University of Chicago Press, Chicago.

Joerges, B. 1988 Large technical systems: Concepts and issues. In *The Development of Large Technical Systems*, edited by R. Mayntz and T. P. Hughes, pp. 9–35. Westview Press, Boulder, Colorado.

Johnson, A. F., and R. H. Thompson 1963 The Ringo site, southeastern Arizona. *American Antiquity* 28: 465–481.

Johnston, G. 1822 Invoice of Sundry Goods and Merchandise to be taken into Lake Superior for the Spring Trade by the Subscriber, Geo. Johnston, St. Mary's Falls, 8 February, 1822. G. Johnston Papers, Mss Jj 37, Box 1, Steere Special Collection, Bayliss Library, Sault Ste. Marie, Michigan.

Johnstone, B. 1990 *Ojibwa Heritage*. University of Nebraska Press, Lincoln.

Joyce, A. A. 2000 The founding of Monte Albán: Sacred propositions and social practices. In *Agency in Archaeology*, edited by M.-A. Dobres and J. E. Robb, pp. 71–91. Routledge, London.

Karmanski, T. J. 1989 *Deep Woods Frontier: A History of Logging in Northern Michigan*. Wayne State University Press, Detroit.

Keller, C. M., and J. D. Keller 1996 *Cognition and Tool Use: The Blacksmith at Work*. Cambridge University Press, Cambridge.

Kelley, J. C. 1991 The known ballcourts of Durango and Zacatecas. In *The Mesoamerican Ballgame*, edited by V. L. Scarborough and D. R. Wilcox, pp. 87–101. University of Arizona Press, Tucson.

Kelly, R. L. 2000 Elements of a behavioral ecological paradigm for the study of hunter-gatherers. In *Social Theory in Archaeology*, edited by M. B. Schiffer, pp. 63–78. University of Utah Press, Salt Lake City, Utah.

Killick, D. 2004 Social constructionist approaches to the study of technology. *World Archaeology* 36: 571–578.

Kingery, W. D. 1993 Technological systems and some implications with regard to continuity and change. In *History from Things: Essays on Material Culture*, edited by S. Lubar and W. D. Kingery, pp. 215–230. Smithsonian Institution Press, Washington D. C.

Kingery, W. D. 2001 The design process as a critical component of the anthropology of technology. In *Anthropological Perspectives on Technology*, edited by M. B. Schiffer, pp. 123–138. University of New Mexico Press, Albuquerque.

Kinzie, J. A. 1932 *Wau-bun, The Early Days in the Northwest*. Lakeside Press, Chicago.

Kite, C. 1788 *An Essay on the Recovery of the Apparently Dead*. London.

Kobayashi, M. 1994 Use-alteration analysis of Kalinga pottery: Interior carbon deposits of cooking pots. In *Kalinga Ethnoarchaeology: Expanding Archaeological Method and Theory*, edited by W. A. Longacre and J. M. Skibo, pp. 127–168. Smithsonian Institution Press, Washington, D. C.

Kopytoff, I. 1986 The cultural biography of things: Commoditization as process. In *The Social Life of Things*, edited by A. Appadurai, pp. 64–91. Cambridge University Press, Cambridge.

Kowalewski, S. A., G. M. Feinman, L. Finster, and R. Blanton 1991 Pre-Hispanic ballcourts from the Valley of Oaxaca, Mexico. In *The Mesoamerican Ballgame*, edited by V. L. Scarborough and D. R. Wilcox, pp. 25–44. University of Arizona Press, Tucson.

Kramer, C. 1982 *Village Ethnoarchaeology: Rural Iran in Archaeological Perspective*. Academic, New York.

Kramer, C. 1997 *Pottery in Rajasthan: Ethnoarchaeology in Two Indian Cities*. Smithsonian Institution Press, Washington, D. C.

Kuhn, S. L., and C. Sarther 2000 Food, lies, and paleoanthropology: Social theory and the evolution of sharing in humans. In *Social Theory in Archaeology*, edited by M. B. Schiffer, pp. 79–96. University of Utah Press, Salt Lake City, Utah.

LaMotta, V. M. 1999 Formation processes of house floor assemblages. In *The Archaeology of Household Activities*, edited by P. Allison, pp. 19–29. Routledge, London.

LaMotta, V. M., and M. B. Schiffer 2001 Behavioral archaeology: Toward a new synthesis. In *Archaeological Theory Today*, edited by I. Hodder, pp. 14–64. Polity Press, Cambridge.

LaMotta, V. M., and M. B. Schiffer 2005 Archaeological formation processes. In *Archaeology: The Key Concepts*, edited by C. Renfrew and P. Bahn, pp. 121–127. Routledge, London.

Landriani, M. 1784 dell Utilitá dei Conduttori Elettriti. Milano.

Lane, T. 1767 Description of an electrometer invented by Mr. Lane; with an account of some experiments made by him with it. *Philosophical Transactions of the Royal Society* 57: 451–460.

Langdon, W. E. 1877 *The Application of Electricity to Railway Working*. MacMillan, London.

Langenbucher, J. 1780 *Beschreibung einer Beträchtlish Verbesserten Electrisier-Maschine*. Augsburg.

Latour, B. 2002 Ethnography of a "high-tech" case: About Aramis. In *Technological Choices: Transformation in Material Cultures since the Neolithic,* edited by P. Lemonnier, pp. 372–398. Routledge, London.

LeBlanc, S. A. 1982 The advent of pottery in the American southwest. In *Southwestern Ceramics: A Comparative Review*, edited by A. Schroeder, Vol. 15, pp. 129–148. The Arizona Archaeologist. School of American Research, Santa Fe, New Mexico.

Lekson, S. 1999 *Chaco Meridian: Centers of Political Power in the Ancient Southwest*. Altamira Press, Walnut Creek, California.

Lemonnier, P. 1986 The study of material culture today: Toward an anthropology of technical systems. *Journal of Anthropological Archaeology* 5: 147–186.

Lemonnier, P. 1992 *Elements for an Anthropology of Technology*. Anthropological Papers No. 88, Museum of Anthropology, University of Michigan, Ann Arbor.

Lemonnier, P. 2002a Introduction. In *Technological Choices: Transformation in Material Cultures since the Neolithic*, edited by P. Lemonnier, pp. 1–35. Routledge, London.

Lemonnier, P. 2002b *Technological Choices: Transformation in Material Cultures since the Neolithic*. Routledge, London.

Lemonnier, P. 2002c Pigs as ordinary wealth: Technical logic, exchange and leadership in New Guinea. In *Technological Choices: Transformation in Material Cultures since the Neolithic*, edited by P. Lemonnier, pp. 126–156. Routledge, London.

Leonard, R. D., and G. T. Jones 1987 Elements of an inclusive evolutionary model for archaeology. *Journal of Anthropological Archaeology* 6: 199–219.

Leroi-Gourhan, A. 1943 *Evolution et Techniques. L'homme et la Matiére*. Albin Michel, Paris.

Leroi-Gourhan, A. 1945 *Evolution et Techniques. Milieu et Techniques*. Albin Michel, Paris.

Levere, T. H. 1969 Martinus van Marum and the introduction of Lavoisier's chemistry into the Netherlands. In *Martinus van Marum: Life and Work*, edited by R. J. Forbes, Vol. 1, pp. 158–286. H. D. Tjeenk Willink and Zoon, Haarlem, Holland.

Leyenaar, T. J. J. 1992 *Ulama*. The survival of the Mesoamerican ballgame *Ullamaliztli*. *Kiva* 58: 115–154.

Lipe, W. D., and M. Hegmon (editors) 1989 The Architecture of Social Integration in Prehistoric Pueblos. *Crow Canyon Archaeological Center, Occasional Paper, 1*.

Longacre, W. A. 1970 *Archaeology as Anthropology: A Case Study*. Anthropological Papers No. 17. University of Arizona Press, Tucson.

Longacre, W. A. 1985 Pottery use-life among the Kalinga, Northern Luzon, the Philippines. In *Decoding Prehistoric Ceramics*, edited by B. A. Nelson, pp. 334–346. Southern Illinois University Press, Carbondale.

Longacre, W. A. 1991 Sources of ceramic variability among the Kalinga of Northern Luzon. In *Ceramic Ethnoarchaeology*, edited by W. A. Longacre, pp. 95–111. University of Arizona Press, Tucson.

Longacre, W. A. 1995 Why did they invent pottery anyway? In *The Emergence of Pottery: Technology and Innovation in Ancient Societies*, edited by W. K. Barnett and J. W. Hoopes, pp. 277–280. Smithsonian Institution Press, Washington, D. C.

Longacre, W. A. 1999 Standardization and specialization: What's the link? In *Pottery and People: A Dynamic Interaction*, edited by J. M. Skibo and G. M. Feinman, pp. 44–58. University of Utah Press, Salt Lake City, Utah.

Longacre, W. A., J. Xia, and T. Yang 1999 I want to buy a black pot. *Journal of Archaeological Method and Theory* 7: 273–293.

Longfellow, H. W. 2000 *The Song of Hiawatha*. Orion Publishing, London.

Lovett, R. 1756 *The Subtil Medium Prov'd: Or, That Wonderful Power of Nature, So Long Ago Conjectur'd by the Most Ancient and Remarkable Philosophers, Which They Call'd Sometimes Aether but Oftener Elementary Fire, Verify'd*. London.

Lowndes, F. 1787 *Observations on Medical Electricity, Containing a Synopsis of All the Diseases in which Electricity has been Recommended or Applied with Success*. London.

Lyman, R. L., and M. J. O'Brien 2000 Measuring and explaining change in artifact variation with clade-diversity diagrams. *Journal of Anthropological Archaeology* 19: 39–74.

Mahias, M. C. 2002 Pottery techniques in India: Technical variants and social choice. In *Technological Choices: Transformation in Material Cultures since the Neolithic,* edited by P. Lemonnier, pp. 157–180. Routledge, London.

Majewski, T., and M. B. Schiffer 2001 Beyond consumption: Toward an archaeology of consumerism. In *Archaeologies of the Contemporary Past*, edited by V. Buchli and G. Lucas, pp. 26–50. Routledge, London.

Martin, P. S., and J. B. Rinaldo 1960 Excavations in the Upper Little Colorado drainage, eastern Arizona. *Fieldiana Anthropology* 51(1): 1–127.

Martin, S. R. 1989 A reconsideration of aboriginal fishing strategies in the northern Great Lakes region. *American Antiquity* 54: 594–604.

Martin, S. R. 1999 A site for all seasons: Some aspects of life in the upper Peninsula during Late Woodland times. In *Retrieving Michigan's Buried Past: The Archaeology of the Great Lakes State*, edited by J. R. Halsey, pp. 221–227. Canbrook Institute of Science. Bulletin 64, Bloomfield Hills, Michigan.

Mascart, M. E. 1876 *Traité d'Électricité Statique. 2 Vols. Paris: Libraire de L'Académie de Médécine*.

Mason, P. P. 1997 *Schoolcraft's Ojibwa Lodge Stories: Life on the Lake Superior Frontier.* Michigan State University Press, Lansing.
Matkin, G. W. 1990 *Technology Transfer and the University*. American Council on Education, New York.
Matson, R. G. 1991 *The Origins of Southwestern Agriculture*. University of Arizona Press, Tucson.
Mauduyt, P. 1784 *Mémoire sur les Différentes Manieres d'Administer l'Électricité, et Observations sur les Effets qu'elles ont Produits*. The Royal Press, Paris.
Mauss, M. 1968 Les techniques du corps. In *Sociologie et Anthropologie*, pp. 365–386. Presses Universitaires de France, Paris.
McClellan, J. E., III 1985 *Science Reorganized: Scientific Societies in the Eighteenth Century*. Columbia University Press, New York.
McCracken, G. 1988 *Culture and Consumption*. Indiana University Press, Bloomington.
McGee, W. J. 1971 *The Seri Indians of Bahia Kino and Sonora, Mexico*. Rio Grande Press, Glorieta, New Mexico.
McGuire, R. H. 1992 *A Marxist Archaeology*. Academic, New York.
McGuire, R. H., and M. B. Schiffer 1983 A theory of architectural design. *Journal of Anthropological Archaeology* 2: 277–303.
McKenney, T. L. 1959 *Sketches of a Tour to the Lakes, of the Character and Customs of the Chippewa Indians, and of Incidents Connected with the Treaty of Fond Du Lac*. Ross and Haines, Minneapolis.
McPherron, A. L. 1967 *The Juntunen Site and Late Woodland Prehistory of the Upper Great Lakes Area*. Anthropological Papers No. 30, Museum of Anthropology, University of Michigan, Ann Arbor.
Mera, H. P. 1934 *Observations on the Archaeology of the Petrified Forest National Monument*. New Mexico Laboratory of Anthropology Technical Series Bulletin 7. Santa Fe.
Meyer, H. W. 1972 *A History of Electricity and Magnetism*. Burndy Library, Norwalk, Connecticut.
Michigan Pioneer and Historical Society 1985 *Proceedings of a Board of Survey Held at Drummond Island, 1822*. In *Historical Collections*, Vol. 23, pp. 202–212. Robert Smith and Co., Lansing.
Millburn, J. R. 1976 *Benjamin Martin: Author, Instrument-Maker, and "Country Showman."* Noordhoff, Leyden.
Miller, D. 1987 *Material Culture and Mass Consumption*. Basil Blackwell, Oxford.
Miller, D. 1998 Why some things matter. In *Material Cultures: Why Some Things Matter*, edited by D. Miller, pp. 3–21. University of Chicago Press, Chicago.
Mills, B., and P. Crown (editors) 1995 *Ceramic Production in the American Southwest*. University of Arizona Press, Tucson.
Mithen, S. (editor) 1998 *Creativity in Human Evolution and Prehistory*. Routledge, London.
Mokyr, J. 1990 *The Lever of Riches: Technological Creativity and Economic Progress*. Oxford University Press, Oxford.
Mom, G. 2004 *The Electric Vehicle: Technology and Expectations in the Automobile Age*. Johns Hopkins University Press, Baltimore.
Morris, E. A. 1980 *Basketmaker Caves in the Prayer Rock District, Northeastern Arizona*. Anthropological Papers of the University of Arizona, No. 35. University of Arizona Press, Tucson.
Morris, E. H. 1927 *The Beginnings of Pottery Making in the San Juan Area: Unfired Prototypes and the Wares of the Earliest Ceramic Period*. Anthropological Papers, Vol. 28, pp. 127–198. American Museum of Natural History, New York.
Morse, E. L. (editor) 1973 *Samuel F. B. Morse: His Letters and Journals*. Da Capo Press, New York.
Munson, P. J. 1969 Comments on Binford's smudge pits and hide smoking: The use of analogy in archaeological reasoning. *American Antiquity* 34: 83–85.
Nairne, E. 1773 Directions for Using the Electrical Machine, as Made and Sold by Edward Naime, Optical, Philosophical, and Mathematical Instrument Maker. London.

Nairne, E. 1774 Electrical Experiments Made with a Machine of his own workmanship, a Description of which is Prefixed. *Philosophical Transactions of the Royal Society* 64: 79–89.

Nairne, E. 1783 The Description and Use of Nairne's Patent Electrical Machine; with the Addition of some Philosophical Experiments and Medical Observations. London.

Nairne, E. 1793 The Description and Use of Naime's Patent Electrical Machine; with the Addition of some Philosophical Experiments and Medical Observations. 4th edition. London.

Neale, J. 1747 *Directions for Gentlemen, who have Electrical Machines, how to Proceed in Making their Experiments*. London.

Neff, H. 1992 Ceramics and evolution. *Archaeological Method and Theory* 4: 141–193.

Neff, H., and D. O. Larson 1997 Methodology of comparison in evolutionary archaeology. In *Rediscovering Darwin: Evolutionary Theory and Archaeological Interpretation*, edited by C. M. Barton and G. A. Clark, pp. 75–94. Archeological Papers No. 7. American Anthropological Association, Washington, D. C.

Neiman, F. D. 1995 Stylistic variation in evolutionary perspective: Inferences from decorative diversity and interassemblage distance in Illinois Woodland ceramic assemblages. *American Antiquity* 60: 7–36.

Nelson, M. C. 1991 The study of technological organization. In *Archaeological Method and Theory*, edited by M. B. Schiffer, Vol. 3. pp. 57–100. University of Arizona Press, Tucson.

Newell, G. E., and E. Gallaga 2004 *Surveying the Archaeology of Northwest Mexico*. University of Utah Press, Salt Lake City, Utah.

Nicholson, W. 1795 *A Dictionary of Chemistry*. 2 Vols. London.

Nickles, D. P. 2004 *Under the Wire: How the Telegraph Changed Diplomacy*. Harvard University Press, Cambridge.

Nielsen, A. E. 1991 Trampling the archaeological record: An experimental study. *American Antiquity* 56: 483–503.

Nielsen, A. E. 1995 Architectural performance and the reproduction of social power. In *Expanding Archaeology*, edited by J. M. Skibo, W. H. Walker, and A. E. Nielsen, pp. 47–66. University of Utah Press, Salt Lake City, Utah.

O'Brien, M. J. 2005 Evolutionism and North America's archaeological record. *World Archaeology* 37: 26–45.

O'Brien, M. J., and T. D. Holland 1992 The role of adaptation in archaeological explanation. *American Antiquity* 57: 3–59.

O'Brien, M. J., and T. D. Holland 1995 The nature and premise of a selection-based archaeology. In *Evolutionary Archaeology: Methodological Issues*, edited by P. A. Teltser, pp. 175–200. University of Arizona Press, Tucson.

O'Brien, M. J., T. D. Holland, R. J. Hoard, and G. L. Fox 1994 Evolutionary implications of design and performance characteristics of prehistoric pottery. *Journal of Archaeological Method and Theory* 1: 259–304.

O'Brien, M. J., and R. L. Lyman 2000 *Applying Evolutionary Archaeology: A Systematic Approach*. Kluwer Academic/Plenum, New York.

O'Brien, M. J., and R. L. Lyman 2003a *Cladistics in Archaeology*. University of Utah Press, Salt Lake City, Utah.

O'Brien, M. J., and R. L. Lyman 2003b *Style, Function, Transmission: Evolutionary Archaeological Perspectives*. University of Utah Press, Salt Lake City. Utah.

O'Brien, M. J., R. L Lyman, and R. D. Leonard, 1998 Basic incompatibilities between evolutionary and behavioral archaeology. *American Antiquity* 63: 485–498.

O'Brien, M. J., R. L. Lyman, and M. B. Schiffer 2005 *Archaeology as a Process: Processualism and its Progeny*. University of Utah Press, Salt Lake City, Utah.

O'Connell, J. F. 1995 Ethnoarchaeology needs a general theory of human behavior. *Journal of Archaeological Research* 3: 504–255.

Orser, C. E. 1996 *A Historical Archaeology of the Modern World*. Plenum, New York.

Orser, C. E. 2004 *Race and Practice in Archaeological Interpretation*. University of Pennsylvania Press, Philadelphia.

Orser, C. E. 2007 *The Archaeology of Race and Racialization in Historic America*. University Press of Florida, Gainesville.

Ortiz, A. 1972 *The Tewa World*. University of Chicago Press, Chicago, Illinois.

Oswalt, W. 1976 *An Anthropological Analysis of Food-Getting Technology*. Wiley, New York.

Oudshoorn, N., and T. Pinch (editors) 2003 *How Users Matter: the Co-construction of Users and Technology*. MIT, Cambridge.

Oyuela-Caycedo, A. 1995 Rock versus clay: The evolution of pottery technology in the case of San Jacinto I, Columbia. In *The Emergence of Pottery: Technology and Innovation in Ancient Societies*, edited by W. K. Barnett and J. W. Hoopes, pp. 133–144. Smithsonian Institution Press, Washington, D. C.

Parsons, E. C. 1996 [1939] *Pueblo Indian Religion*. University of Nebraska Press, Lincoln.

Pauketat, T. R. 2000 The tragedy of the commoners. In *Agency in Archaeology*, edited by M.-A. Dobres and J. E. Robb, pp. 113–129. Routledge, London.

Pauketat, T. R. 2001 Practice and history in archaeology: An emerging paradigm. *Anthropological Theory* 1: 73–98.

Pauketat, T. R. 2003 Materiality and the immaterial in historical-processual archaeology. In *Essential Tensions in Archaeological Method and Theory*, edited by T. L. VanPool, and C. S. VanPool, pp. 41–54. University of Utah Press, Salt Lake City, Utah.

Pauketat, T. R., and S. M. Alt 2005 Agency in a postmold? Physicality and the archaeology of culture-making. *Journal of Archaeological Method and Theory* 12: 213–236.

Pavlů, I. 1997 *Pottery Origins: Initial Forms, Cultural Behaviour and Decorative Style*. Karolinum, Vydavatelství Univerzity Karlovy, Praha.

Peake, O. B. 1954 *History of the United States Indian Factory System, 1795–1822*. Sage Books, Denver.

Pellegram, A. 1998 The message in paper. In *Material Cultures: Why Some Things Matter*, edited by D. Miller, pp. 103–120. University of Chicago Press, Chicago.

Pfaffenberger, B. 1992 Social anthropology of technology. *Annual Review of Anthropology* 21: 491–516.

Pfaffenberger, B. 2001 Symbols do not create meaning – activities do: Or why symbolic anthropology needs an anthropology of technology. In *Anthropological Perspectives on Technology*, edited by M. B. Schiffer, pp. 215–235. University of New Mexico Press, Albuquerque.

Pierce, C. 2005 Reverse engineering the ceramic pot: Cost and performance properties of plain and textured ceramics. *Journal of Archaeological Method and Theory* 12: 117–157.

Post, R. C. 1976 *Physics, Patents, and Politics: A Biography of Charles Grafton Page*. Science History Publications, New York.

Preece, W. H., and J. Sivewright 1891 *Telegraphy*. 9th edition. Longman, London.

Prescott, G. B. 1888 *Electricity and the Electric Telegraph*. 7th edition. D. Appleton, New York.

Priestley, J. 1779 *Experiments and Observations Relating to the Various Branches of Natural Philosophy*; With a Continuation of the Observations on Air. London.

Priestley, J. 1781 *Experiments and Observations on Different Kinds of Air. 3rd edition*. 3 vols. London.

Priestley, J. 1966(1775) *The History and Present State of Electricity, with Original Experiments*. Johnson Reprint Co., New York.

Quilici-Pacaud, J.F. 2002 Dominant representations and technical choices: A method of analysis with examples from aeronautics. In *Technological Choices: Transformation in Material Cultures since the Neolithic*, edited by P. Lemonnier, pp. 399–412. Routledge, London.

Quill, J. 1985 *Spitfire. A Test-Pilot's Story*. Arrow Books, London.

Rathje, W. L., and C. Murphy 1992 *Rubbish! The Archaeology of Garbage*. Harper Collins, New York.

Rathje, W. L., and M. B. Schiffer 1982 *Archaeology*. Harcourt Brace Jovanovich, New York.

Ravesloot, J. C. 1979 The Animas Phase: Post-Classic Mimbres occupation of the Mimbres Valley, New Mexico. M. A. Thesis, Department of Anthropology, Southern Illinois University, Illinois.

Reid, J. D. 1879 *Telegraph in America. Its Founders Promoters and Noted Men*. Derby Brothers, New York.
Reid, J. J., W. L. Rathje, and M. B. Schiffer 1974 Expanding archaeology. *American Antiquity* 39: 125–126.
Reid, J. J., M. B. Schiffer, and W. L. Rathje 1975 Behavioral archaeology: Four strategies. *American Anthropologist* 77: 864–868.
Reid, K. C. 1984 Fire and ice: New evidence for the production and preservation of Late Archaic fiber-tempered pottery in the middle-latitude lowlands. *American Antiquity* 49: 55–76.
Reid, K. C. 1990 Simmering down: A second look at Ralph Linton's "North American cooking pots." In *Hunter-Gatherer Pottery in the Far West*, edited by D. R. Tuohy and A. J. Dansie, pp. 7–17. Anthropology Papers No. 23, Nevada State Museum, Carson City.
Reina, R. E., and R. M. Hill II 1978 *The Traditional Pottery of Guatemala*, University of Texas Press, Austin.
Rice, P. M. 1987 *Pottery Analysis: A Sourcebook*. University of Chicago Press, Chicago.
Rice, P. M. 1996a Recent ceramic analysis: 1. Function, style, and origins. *Journal of Archaeological Research* 4: 133–163.
Rice, P. M. 1996b Recent ceramic analysis: 2. Composition, production, and theory. *Journal of Archaeological Research* 4: 165–202.
Rice, P. M. 1999 On the Origins of Pottery. *Journal of Archaeological Method and Theory* 6: 1–54.
Richards, M. 1966 Brains, bones, and hotsprings: Native American deerskin dressing at the time of contact. *Bulletin of Primitive Technology* 12, Fall 1966.
Roberts, P. A. 1988 *Technology Transfer: A Policy Model*. National Defense University Press, Washington, D. C.
Ritzenthaler, R. E 1949 The Chippewa Indian method of securing and tanning deerskin. *Wisconsin Archaeologist* 28: 6–13.
Ritzenthaler, R. E., and P. Ritzenthaler 1983 *The Woodland Indians of the Western Great Lakes*. Milwaukee Public Museum, Milwaukee.
Roberts, N. A. 1991 *Cultural Resources Overview and National Register Evaluation of Historic Structures, Grand Island National Recreation Area, Michigan*. USDA-Forest Service, Hiawatha National Forest, Escanaba, Michigan. RN Cultural Resource Consultants, Minneapolis, Minnesota.
Roux, V. 2003 A dynamic systems framework for studying technological change: Application to the emergence of the potter's wheel in the southern Levant. *Journal of Archaeological Method and Theory* 10: 1–30.
Roux, V. 2007 Ethnoarchaeology: A non historical science of reference necessary for interpreting the past. *Journal of Archaeological Method and Theory* 14: 153–178.
Rye, O. S. 1976 Keeping your temper under control. *Archaeology and Physical Anthropology in Oceania* 11: 106–137.
Sabine, R. 1869 *The History and Progress of the Electric Telegraph*. 2nd edition. Van Nostrand, New York.
Sackett, J.R. 1977 The meaning of style in archaeology: A general model. *American Antiquity* 42: 369–380.
Sans, C. 1772 Guérison de la Paralysie, par L'Eléctricité, ou cette Expérience Physique Employée avec Succes dans Le Traitement de cette Maladie Regardée a Présent comme Incurable. Paris.
Santley, R. S., M. J. Berman, and R. T. Alexander 1991 The politicization of the Mesoamerican ballgame and its implications for the interpretation of the distribution of ball courts in central Mexico. In *The Mesoamerican Ballgame*, edited by V. L. Scarborough and D. R. Wilcox, pp. 3–24. University of Arizona Press, Tucson.
Sapir, E. 1923 A note on Sarcee pottery. *American Anthropologist* 25: 247–253.
Sassaman, K. E. 1993 *Early Pottery in the Southeast: Tradition and Innovation in Cooking Technology*. University of Alabama Press, Tucsaloosa.
Sassaman, K. E. 1995 The social contradictions of traditional and innovative cooking technologies in the prehistoric American southeast. In *The Emergence of Pottery: Technology and*

Innovation in Ancient Societies, edited by W. K. Barnett and J. W. Hoopes, pp. 223–240. Smithsonian Institution Press, Washington D. C.

Sassaman, K. E. 2005 Poverty Point as structure, event, process. *Journal of Archaeological Method and Theory* 12: 335–364.

Schaafsma, C. F., and C. L. Riley (editors) 1999 *The Casas Grandes World*. University of Utah Press, Salt Lake City, Utah.

Schaafsma, P. 1997 *Rock Art Sites in Chihuahua Mexico*. Archaeology Notes 171. Office of Archaeological Studies, Museum of New Mexico, Santa Fe.

Schaffer, S. 1983 Natural philosophy and public spectacle in the eighteenth century. *History of Science* 21: 1–43.

Schellen, H. 1850 *Der Electromagnetische Telegraph*. Bieweg, Braunschweig.

Schiffer, M. B. 1972 Archaeological context and systemic context. *American Antiquity* 37: 156–165.

Schiffer, M. B. 1975a Archaeology as behavioral science. *American Anthropologist* 77: 836–848.

Schiffer, M. B. 1975b Behavioral chain analysis. Activities, organization, and the use of space. In *Chapters in the Prehistory of Eastern Arizona IV*, edited by P. S. Martin et al. Field Museum of Natural History, Chicago, Fieldiana: Anthropology 65: 103–119.

Schiffer, M. B. 1976 *Behavioral Archeology*. Academic, New York.

Schiffer, M. B. 1979 *A preliminary consideration of behavioral change. In Transformations*: Mathematical Approaches to Culture Change, edited by C. Renfrew and K. Cooke, pp. 353–368. Academic Press, New York.

Schiffer, M. B. 1983 Toward the identification of formation processes. *American Antiquity* 48: 675–706.

Schiffer, M. B. 1985 Is there a "Pompeii Premise" in archaeology? *Journal of Archaeological Research* 41: 18–41.

Schiffer, M. B. 1987 *Formation Processes of the Archaeological Record*. University of New Mexico Press, Albuquerque.

Schiffer, M. B. 1988a The effects of surface treatment on permeability and evaporative cooling effectiveness of pottery. In *Proceedings of the 26th International Archaeometry Symposium*, edited by R. M. Farquhar, R. G. V. Hancock, and L. A. Pavlish, pp. 23–29. Archaeometry Laboratory, Department of Physics, University of Toronto, Toronto.

Schiffer, M. B. 1988b The structure of archaeological theory. *American Antiquity* 53: 461–485.

Schiffer, M. B. 1989 Formation processes of Broken K Pueblo: Some hypotheses. In *Quantifying Diversity in Archaeology*, edited by R. D. Leonard and G. T. Jones, pp. 37–58. Cambridge University Press, Cambridge.

Schiffer, M. B. 1990 Technological change in water-storage and cooking pots: Some predictions from experiments. In *The Changing Roles of Ceramics in Society: 26,000 B. P. to the Present*, edited by W. D. Kingery, pp. 119–136. The American Ceramic Society, Westerville, Ohio.

Schiffer, M. B. 1991 *The Portable Radio in American Life*. University of Arizona Press, Tucson.

Schiffer, M. B. 1992 *Technological Perspectives on Behavioral Change*. University of Arizona Press, Tucson.

Schiffer, M. B. 1993 Cultural imperatives and product development: The case of the shirt-pocket radio. *Technology and Culture* 34: 98–113.

Schiffer, M. B. 1995a *Behavioral Archaeology: First Principles*. University of Utah Press, Salt Lake City, Utah.

Schiffer, M. B. 1995b Social theory and history In behavioral archaeology. In *Expanding Archaeology*, edited by J. M. Skibo, W. H. Walker, and A. E. Nielsen, pp. 22–35. University of Utah Press, Salt Lake City, Utah.

Schiffer, M. B. 1996 Some relationships between behavioral and evolutionary archaeologies. *American Antiquity* 61: 643–662.

Schiffer, M. B. 2000 Indigenous theories, scientific theories, and product histories. In *Matter, Materiality, and Modem Culture*, edited by P. Graves-Brown, pp. 72–96. Routledge, London.

Schiffer, M. B. 2001a The explanation of long-term technological change. In *Anthropological Perspectives on Technology*, edited by M. B. Schiffer, pp. 215–235. University of New Mexico Press, Albuquerque.

Schiffer, M. B. 2001b (editor) *Anthropological Perspectives on Technology*. University of New Mexico Press, Albuquerque.

Schiffer, M. B. 2002 Studying technological differentiation: The case of 18th-century electrical technology. *American Anthropologist* 104: 1148–1161.

Schiffer, M. B. 2004 Studying technological change: A behavioral perspective. *World Archaeology* 36: 579–585.

Schiffer, M. B. 2005a The devil is in the details: The cascade model of invention processes. *American Antiquity* 485–502.

Schiffer, M. B. 2005b The electric lighthouse in the nineteenth century. *Technology and Culture* 46: 275–305.

Schiffer, M. B. 2007 Some thoughts on the archaeological study of social organization. In *Archaeological Anthropology: Perspectives on Method and Theory*, edited by J. M. Skibo, M. W. Graves, and M. T. Stark, pp. 57–71. University of Arizona Press, Tucson.

Schiffer, M. B. 2008 Behavioral archaeology. In *Encyclopedia of Archaeology*, edited by D. Pearsall, Vol.2, pp. 909–919 Elsevier, New York.

Schiffer, M. B., T. C. Butts, and K. K. Grimm 1994 *Taking Charge: The Electric Automobile in America*. Smithsonian Institution Press, Washington, D. C.

Schiffer, M. B., T. E. Downing, and M. McCarthy 1981 Waste not, want not: An ethnoarchaeological study of reuse in Tucson, Arizona. In *Modern Material Culture: The Archaeology of Us*, edited by R. A. Gould and M. B. Schiffer, pp. 68–86. Academic, New York.

Schiffer, M. B., K. L. Hollenback, and C. L. Bell 2003 *Draw the Lightening Down: Benjamin Franklin and Electrical Technology in the Age of Enlightenment*. University of California Press, Berkeley.

Schiffer, M. B., and J. H. House (assemblers) 1975 *The Cache River Archeological Project: An Experiment in Contract Archaeology*. Arkansas Archeological Survey, Research Series No. 8.

Schiffer, M. B., and A. R. Miller 1999a *The Material Life of Human Beings: Artifacts, Behavior, and Communication*. Routledge, London.

Schiffer, M. B., and A. R. Miller 1999b A behavioral theory of meaning. In *Pottery and People: A Dynamic Interaction*, edited by J. M. Skibo and G. Feinman, pp. 199–217. University of Utah Press, Salt Lake City, Utah.

Schiffer, M. B., and J. M. Skibo 1987 Theory and experiment in the study of technological change. *Current Anthropology* 28: 595–622.

Schiffer, M. B., and J. M. Skibo 1997 The explanation of artifact variability. *American Antiquity* 62: 27–50.

Schiffer, M. B., J. M. Skibo, T. C. Boelke, M. A. Neupert, and M. Aronson 1994 New perspectives on experimental archaeology: Surface treatments and thermal response of the clay cooking pot. *American Antiquity* 59: 197–217.

Schiffer, M. B., J. M. Skibo, J. M. Griffiths, K. L. Hollenback, and W. A. Longacre 2001 Behavioral archaeology and the study of technology. *American Antiquity* 66: 729–737.

Schoolcraft, H. R. 1821 *Narrative Journal of Travels Through the Northwestern Regions of the United States extending from Detroit through the great chain of American lakes to the sources of the Mississippi River in the year* 1820. E. & E. Hosford, Albany.

Schoolcraft, H. R. 1851 *Personal memoirs of a residence of 30 years with the Indian tribes on the American frontier*. Lippincott, Grambo, Philidelphia.

Schoolcraft, H. R. 1853 Census returns of the Indian tribes of the U.S. with their vital statistics and industrial statistics. In *Indian tribes of the United States*, pp. 458–467. Lippincott, Grambo, Philadelphia.

Schroeder, A. H. 1982 Historical review of Southwestern ceramics. Southwestern ceramics: A comparative review, In *The Arizona Archaeologist*, Vol. 15, edited by A. H. Schroeder, pp. 1–26. School of American Research, Santa Fe, New Mexico.

Shackel, P. 1996 *Culture Change and the New Technology: An Archaeology of the Early Industrial Era*. Plenum, New York.
Shaffner, T. P. 1859 *The Telegraph Manual*. Pudney & Russell, New York.
Shanks, M., and C. Tilley 1997 *Re-Constructing Archaeology: Theory and Practice*. 2nd edition. Routledge, London.
Shepard, A. O. 1953 Notes on color and paste composition. In *Archaeological Studies in the Petrified Forest National Monument*, edited by F. Wendorf, pp. 177–193. Museum of Northern Arizona Bulletin 27, Flagstaff.
Shimada, I., and U. Wagner 2007 A holistic approach to pre-Hispanic craft production. In *Archaeological Anthropology: Perspectives on Method and Theory*, edited by J. M. Skibo, M. W. Graves, and M. T. Stark, pp. 163–197. University of Arizona Press, Tucson.
Shott, M. J. 1997 Innovation and selection in prehistory: A case study from the American bottom. In *Stone Tools: Theoretical Insights into Human Prehistory*, edited by G. H. Odell, pp. 279–314. Plenum, New York.
Sigaud de la Fond, J. 1776 Traité De i'Électricité, dans Lequel on Expose & on Démontre par Expérience, toutes les Déscouvertes Electriques, Faites Jusqu'á ce Jour. Paris.
Sillar, B., and M. S. Tite 2000 The challenge of 'technological choices' for materials science approaches in archaeology. *Archaeometry* 42: 2–20.
Silva, F. A. 2008 Ceramic technology of the *Asurini do Xingu*, Brazil: An ethnoarchaeological study of artifact variability. *Journal of Archaeological Method and Theory*.
Skibo, J. M. 1992 *Pottery Function: A Use-Alteration Perspective*. Plenum, New York.
Skibo, J. M. 1994 The Kalinga cooking pot: An ethnoarchaeological and experimental study of technological change. In *Kalinga Ethnoarchaeology: Expanding Archaeological Method and Theory*, edited by W. A. Longacre and J. M. Skibo, pp. 113–126. Smithsonian Institution Press, Washington, D. C.
Skibo, J. M., and E. Blinman 1999 Exploring the origins of pottery on the Colorado Plateau. In *Pottery and People: A Dynamic Interaction*, edited by J. M. Skibo and G. M. Feinman, pp. 171–183. University of Utah Press, Salt Lake City, Utah.
Skibo, J. M., and G. Feinman (editors) 1999 *Pottery and People:* A Dynamic Interaction. University of Utah Press, Salt Lake City, Utah.
Skibo, J. M., T. J. Martin, E. C. Drake, and J. G. Franzen 2004 *Gete Odena*: Grand Island's post-contact occupation at William's landing. *Midcontinental Journal of Archaeology* 29: 167–189.
Skibo, J. M., E. B. McCluney, and W. H. Walker (editors) 2002 *The Joyce Well Site: On the Frontier of the Casas Grandes World*. University of Utah Press, Salt Lake City, Utah.
Skibo, J. M., and M. B. Schiffer 1995 The clay cooking pot: An exploration of women's technology. In *Expanding Archaeology*, edited by J. M. Skibo, W. H. Walker, and A. E. Nielsen, pp. 80–91. University of Utah Press, Salt Lake City, Utah.
Skibo, J. M., and M. B. Schiffer 2001 Understanding artifact variability and change: A behavioral framework. In *Anthropological Perspectives on Technology*, edited by M. B. Schiffer, pp. 139–149. University of New Mexico Press, Albuquerque.
Skibo, J. M., M. B. Schiffer, and N. Kowalski 1989 Ceramic style analysis in archaeology and ethnoarchaeology: Bridging the analytical gap. *Journal of Anthropological Archaeology* 83: 388–409.
Skibo, J. M., M. B. Schiffer, and K. C. Reid 1989 Organic-tempered pottery: An experimental study. *American Antiquity* 54: 122–146.
Skibo, J. M., and W. H. Walker 2002 Ball courts and ritual performance. In *The Joyce Well Site: On the Frontier of the Casas Grandes World*, edited by J. M. Skibo, E. McCluney, and W. H. Walker, pp. 107–128. University of Utah Press, Salt Lake City, Utah.
Skibo, J. M., W. H. Walker, and A. E. Nielsen 1995 *Expanding Archaeology*. University of Utah Press, Salt Lake City, Utah.
Smith, M. L. 2007 Inconspicuous consumption: Non-display goods and identity formation. *Journal of Archaeological Method and Theory* 14:421–438.
Smith, R. A. 1972 *A Social History of the Bicycle, Its Early Life and Times in America*. American Heritage Press, New York.

South, S. 1977 *Method and Theory in Historical Archeology*. Academic, New York.

Spaulding, A. C. 1968 Explanation in archeology. In *New Perspectives in Archaeology*, edited by S. R. Binford and L. R. Binford, pp. 33–39. Aldine, Chicago.

Spector, J. D. 1975 An historic Winnebago Indian site in Jefferson County, Wisconsin. *The Wisconsin Archaeologist* 56: 270–345.

Spencer, C. S. 1997 Evolutionary approaches in archaeology. *Journal of Archaeological Research* 5: 209–264.

Spencer-Wood, S. M. (editor) 1987 *Consumer Choice in Historical Archaeology*. Plenum, New York.

Spurr, K., and K. Hays-Gilpin 1996 New evidence for early Basketmaker III ceramics from the Kayenta area. Paper presented at the 69th annual Pecos Conference, Flagstaff, Arizona.

Stahl, A. B. 1989 Plant-food processing: Implications for dietary quality. In *Foraging and Farming: The Evolution of Plant Exploitation*, edited by D. R. Harris and G. C. Hillman, pp. 171–194. Unwin Hyman, New York.

Stark, M. T. 1994 Pottery exchange and the regional system: A Dalupa case study. In *Kalinga Ethnoarchaeology: Expanding Archaeological Method and Theory,* edited by W. A. Longacre and J. M. Skibo, pp. 169–198. Smithsonian Institution Press, Washington, D. C.

Stark, M. T. 1998 Technical choices and social boundaries in material culture patterning: An introduction. In *The Archaeology of Social Boundaries*, edited by M. T. Stark, pp. 1–11. Smithsonian Institution Press, Washington, D. C.

Stark, M. T. 2003 Current issues in ceramic ethnoarchaeology. *Journal of Archaeological Research* 11: 193–242.

Stark, M. T., and J. M. Skibo 2007 A history of the Kalinga ethnoarchaeology project. In *Archaeological Anthropology: Perspectives on Method and Theory,* edited by J. M. Skibo, M. W. Graves, and M. T. Stark, pp. 93–110. University of Arizona Press, Tucson.

Staudenmaier, J. M. 1985 *Technology's Storytellers: Reweaving the Human Fabric*. Society for the History of Technology and MIT, Cambridge.

Staudenmaier, J. M. 2002 Rationality, agency, contingency: Recent trends in the history of technology. *Reviews in American History* 30: 168–181.

Sullivan, A. P., III 2007 Archaeological anthropology and strategies of knowledge formation in American archaeology. In *Archaeological Anthropology: Perspectives on Method and Theory*, edited by J. M. Skibo, M. W. Graves, and M. T. Stark, pp. 40–56. University of Arizona Press, Tucson.

Sullivan, A. P. III, and K. Rozen 1985 Debitage analysis and archaelogical interpretation. American Antiquity 50:755–779.

Sutphen, H. R. 1901 Touring in automobiles. *Outing* 38: 197–202.

Sutton, G. V. 1995 *Science for a Polite Society: Gender, Culture, and the Demonstration of Enlightenment*. Westview Press, Boulder.

Tagg, M. D. 1996 Early cultigens from Fresnal shelter, southeastern New Mexico. *American Antiquity* 61: 311–324.

Taylor, W. B. 1879 An Historical Sketch of Henry's Contribution to the Electro-Magnetic Telegraph: With an Account of the Origin and Development of Prof. Morse's Invention. *Annual Report of the Smithsonian Institution for 1878*. U.S. Government Printing Office, Washington, D. C.

Teltser, P. A. (editor) 1995 *Evolutionary Archaeology: Methodological Issues*. University of Arizona Press, Tuscon.

Thomas, N. 1989 *Entangled Objects: Exchange, Material Culture, and Colonialism in the Pacific*. Harvard University Press, Cambridge, Massachusetts.

Thompson, R. A., and G. B. Thompson 1974 *Preliminary Report of Excavations in the Grand Canyon National Monument, Sites: GC-670, GC-663*. Report prepared for the National Park Service, Southern Utah State College.

Thwaites, R. G. 1910 *A Wisconsin Fur Traders Journal, The Fur Trade in the Upper Lakes – 1778–1815, and The Fur Trade in Wisconsin – 1815–1817*. Collections of the State Historical Society of Wisconsin, Vol. 19. Madison.

Toaldo, D. G. 1779 *Mémoires sur les Conducteurs pour Préserver les Edificés de la Foudre*. Strasbourg.
Trostel, B. 1994 Household pots and possessions: An ethnoarchaeological study of material goods and wealth. In *Kalinga Ethnoarchaeology: Expanding Archaeological Method and Theory*, edited by W. A. Longacre and J. M. Skibo, pp. 209–224. Smithsonian Institution Press, Washington, D. C.
Tuohy, D. R., and A. J. Dansie (editors) 1990 *Hunter-Gatherer Pottery in the Far West*. Anthropology Papers No. 23, Nevada State Museum, Carsen City.
United States Commissioner of Patents 1883 *Subject-Matter Index of Patents for Inventions Granted in France from 1791 to 1876 Inclusive*. U.S. Government Printing Office, Washington, D. C.
Vail, A. 1845 *The American Electro Magnetic Telegraph*. Lea & Blanchard, Philadelphia.
Vail, J. C. (editor) 1914 *Early History of the Electro-Magnetic Telegraph, from the Letters and Journals of Alfred Vail*. Hine Brothers, New York.
van der Leeuw, S. 2002 Giving the potter a choice: Conceptual aspects of pottery techniques. In *Technological Choices: Transformation in Material Cultures since the Neolithic*, edited by P. Lemonnier, pp. 238–288. Routledge, London.
van der Leeuw, S., and R. Torrence (editors) 1989 *What's New? A Closer Look at the Process of Innovation*. Unwin Hyman, London.
Vandiver, P., O. Soffer, B. Klima, and J. Svoboda 1989 The origins of ceramic technology at Dolní Véstonice, Czechoslovakia. *Science* 246: 1002–1008.
VanPool, C. S. 2003 The shaman-priests of the Casas Grandes region, Chihuahua, Mexico. *American Antiquity* 68: 696–717.
VanPool, C. S., and T. L. VanPool 2003a Introduction: Method, theory and the essential tension. In *Essential Tensions in Archaeological Method and Theory*, edited by T. L. VanPool, and C. S. VanPool, pp. 1–4. University of Utah Press, Salt Lake City, Utah.
VanPool, T. L., and C. S. VanPool 2003b Agency and evolution: The role of intended and unintended consequences of action. In *Essential Tensions in Archaeological Method and Theory*, edited by T. L. VanPool, and C. S. VanPool, pp. 89–114. University of Utah Press, Salt Lake City, Utah.
Varien, M. D., and B. J. Mills 1997 Accumulations research: Problems and prospects for estimating site occupation span. *Journal of Archaeological Method and Theory* 4: 141–191.
Vitelli, K. D. 1989 Were pots first made for foods? Doubts from Franchthi. *World Archaeology* 21: 17–29.
Vitelli, K. D. 1995 Pots, potters, and the shaping of Greek Neolithic society. In *The Emergence of Pottery: Technology and Innovation in Ancient Societies*, edited by W. K. Barnett and J. W. Hoopes, pp. 55–63. Smithsonian Institution Press, Washington, D. C.
Vitelli, K. D. 1999 "Looking up" at early ceramics in Greece. In *Pottery and People A Dynamic Interaction*, edited by J. M. Skibo and G. M. Feinman, pp. 184–198. University of Utah Press, Salt Lake City, Utah.
Volta, A. 1800 On the electricity excited by the mere contact of conducting substances of different kinds. *Philosophical Transactions of the Royal Society* 90: 403–431.
Walker, W. H. 1995a Ceremonial trash? In *Expanding Archaeology*, edited by J. M. Skibo, W. H. Walker, and A. E. Nielsen, pp. 67–79. University of Utah Press, Salt Lake City, Utah.
Walker, W. H. 1995b Ritual Prehistory: A Pueblo Case Study. Ph. D. dissertation, University of Arizona. University Microfilms, Ann Arbor.
Walker, W. H. 1996 Ritual deposits: Another perspective. In River of Change: Prehistory of the Middle Colorado River Valley, Arizona, edited by E. C. Adams, pp. 75–91. Arizona State Museum Archaeological Series No. 185, Tucson.
Walker, W. H. 1998 Where are the witches of prehistory? *Journal of Archaeological Method and Theory* 5: 245–308.
Walker, W. H. 2001 Ritual technology in an extranatural world. In *Anthropological Perspectives on Technology*, edited by M. B. Schiffer, pp. 87–106. University of New Mexico Press, Albuquerque.
Walker, W. H., and L. J. Lucero 2000 The depositional history of ritual and power. In *Agency in Archaeology*, edited by M.-A. Dobres and J. E. Robb, pp. 130–147. Routledge, New York, London.

Walker, W. H., and M. B. Schiffer 2006 The materiality of social power: The artifact-acquisition perspective. *Journal of Archaeological Method and Theory* 13: 67–88.

Walker, W. H., and J. M. Skibo 2002 Joyce Well and the Casas Grandes religious interaction sphere. In *The Joyce Well Site: On the Frontier of the Casas Grandes World,* edited by J. M. Skibo, E. B. McCluney, pp. 167–175. University of Utah Press, Salt Lake City, Utah.

Walker, W. H., J. M. Skibo, and A. E. Nielsen 1995 Introduction: Expanding archaeology. In *Expanding Archaeology*, edited by J. M. Skibo, W. H. Walker, and A. E. Nielsen, pp. 1–14. University of Utah Press, Salt Lake City, Utah.

Waselkov, G. A. 1992 French and colonial trade in the Upper Creek Country. In *Calumet and Fleur-de-Lys: Archaeology of Indian and French Contact in the Midcontinent*, edited by J. A. Walthall and T. E. Emerson, pp. 35–54. Smithsonian Institution Press, Washington, D. C.

Webers, J. 1781 *Neue Erfahrungen Idiolektrische Körper ohne Einiges Reiben zu Elektrisiren.* Augsburg.

Weiner, A. B. 1985 Inalienable wealth. *American Ethnologist* 12: 210–227.

Wendorf, F. 1950 The Flattop Site in the Petrified Forest National Monument. *Plateau* 22: 43–51.

Wendorf, F. 1953 *Archaeological Studies in the Petrified Forest National Monument.* Museum of Northern Arizona, Bulletin 27. Flagstaff, Arizona.

Wesley, J. 1792 *Primitive Physic: Or an Easy and Natural Method of Curing Most Diseases.* 24th edition. London.

Whalen, M. E. 1981 Cultural-ecological aspects of the Pithouse-to-Pueblo transition in a portion of the Southwest. *American Antiquity* 46: 75–92.

Whalen, M. E., and P. E. Minnis 1996 Ballcourts and political centralization in the Casas Grandes region. *American Antiquity* 61: 732–746.

Whalen, M. E., and P. E. Minnis 1999 Investigating the Paquimé regional system. In *The Mesoamerican Ballgame*, edited by V. L. Scarborough and D. R. Wilcox, pp. 54–62. University of Arizona Press, Tucson.

Whalen, M. E., and P. E. Minnis 2001 *Casas Grandes and its Hinterland: Prehistoric Regional Organization in Northwest Mexico.* University of Arizona Press, Tucson.

Wilcox, D. R. 1991 The Mesoamerican ballgame in the American Southwest. In *The Mesoamerican Ballgame*, edited by V. L. Scarborough and D. R. Wilcox, pp. 101–125. University of Arizona Press, Tucson.

Wilcox, D. R., and C. Sternberg 1983 *Hohokam Ballcourts and their Interpretation.* Arizona State Museum Archaeological Series No. 160. University of Arizona Press.

Wilk, R. R. 2001 Toward an archaeology of needs. In *Anthropological Perspectives on Technology*, edited by M. B. Schiffer, pp. 107–122. University of New Mexico Press, Albuquerque.

Wilkerson, S. J. 1991 And then they were sacrificed: The ritual ballgame of northeastern Mesoamerica through time and space. In *Mesoamerican Ballgame*, edited by V. L. Scarborough and D. R. Wilcox, pp. 45–72. University of Arizona Press, Tucson.

William, F., and D. V. Gibson 1990 *Technology Transfer: A Communication Perspective.* Sage, Newbury Park, California.

Williams, M. L. 1992 *Schoolcraft's Narrative Journal of Travels Through the Northwestern Regions of the United States Extending from Detroit Through the Great Chain of American Lakes to the Sources of the Mississippi River in the Year 1820.* Michigan State University Press, Lansing.

Wills, W. H. 1988 *Early Prehistoric Agriculture in the American Southwest.* University of Washington Press, Seattle.

Wilson, B. 1773 Observations upon lightning, and the method of securing buildings from its effects. *Philosophical Transactions of the Royal Society* 63: 49–65.

Wilson, C. D. 1989 Sambrito "Brown" from site LA 4169, a description and evaluation. *Pottery Southwest* 6: 4–5.

Wilson, C. D., and E. Blinman 1993 *Upper San Juan Region Pottery Technology.* Office of Archaeological Studies, Archaeology Notes 80. Museum of New Mexico, Santa Fe.

Wilson, C. D., and E. Blinman 1994 Early Anasazi ceramics and the basketmaker tradition. In *Proceedings of the Anasazi Symposium 1991*, compiled by A. Hutchinson and J. E. Smith, pp. 199–211. Mesa Verde Museum Association, Mesa Verde, Colorado.

Wilson, C. D., and E. Blinman 1995 Changing specialization of white ware manufacture in the Northern San Juan region. In *Ceramic Production in the American Southwest*, edited by B. J. Mills and P. L. Crown, pp. 63–87. University of Arizona Press, Tucson.

Wilson, C. D., E. Blinman, J. M. Skibo, and M. B. Schiffer 1996 Designing of Southwestern pottery: A technological and experimental approach. In *Interpreting Southwestern Diversity: Underlying Principles and Overarching Patterns*, edited by P. R. Fish and J. J. Reid, pp. 249–256. Anthropological Papers No. 48, Arizona State Museum, Tempe.

Wobst, H. M. 1977 Stylistic behavior and information exchange. In *Papers for the Director: Essays in Honor of James B. Griffin*, edited by C. E. Cleland, pp. 317–334. Anthropological Papers No. 61, University of Michigan, Museum of Anthropology, Ann Arbor.

Wobst, H. M. 2000 Agency in (spite of) material culture. In *Agency in Archaeology*, edited by M.-A. Dobres and J. E. Robb, pp. 40–50. Routledge, London.

Work, J. 1914 Journal of John Work, Dec. 15th, 1825 to June 12th, 1826 (edited by T. C. Elliot). *Washington Historical Quarterly* 5(4): 258–287.

Wylie, A. 1995 An expanded behavioral archaeology: Transformation and redefinition. In *Expanding Archaeology*, edited by J. M. Skibo, W. H. Walker, and A. E. Nielsen, pp. 198–209. University of Utah Press, Salt Lake City, Utah.

Wylie, A. 2000 Questions of evidence, legitimacy, and the (dis)unity of science. *American Antiquity* 65: 227–238.

Zedeño, M. N. 1997 Landscapes, land use, and the history of territory formation: An example from the Puebloan Southwest. *Journal of Archaeological Method and Theory* 4: 67–103.

Zedeño, M. N. 2000 On what people make of places: A behavioral cartography. In Social Theory in Archaeology, edited by M. B. Schiffer. University of Utah Press, Salt Lake City.

Zipf, G. K. 1941 *Human Behavior and the Principle of Least Effort*. Addison-Wesley, Cambridge, Massachusetts.

Index

Printed in the United States
206004BV00002B/1/P

9 780387 771328